The
Pasteurization
of France

The Pasteurization of France

Bruno Latour

*Translated by Alan Sheridan
and John Law*

Harvard University Press
Cambridge, Massachusetts
and London, England

Originally published as *Les microbes: guerre et paix suivi de irréductions,*
copyright © 1984 Editions A. M. Métailié, Paris.

Translation of this book has been aided by a grant from the
Georges Lurcy Charitable and Educational Trust.

First Harvard University Press paperback edition, 1993

Library of Congress Cataloging in Publication Data

Latour, Bruno.
  [Microbes. English]
  The pasteurization of France / Bruno Latour ; translated by Alan Sheridan
and John Law.
  p. cm.
  Rev. translation of: Les microbes : guerre et paix ; suivi de,
Irréductions.
  Bibliography: p.
  Includes index.
  Contents: War and peace of microbes—Irreductions.
  ISBN 0-674-65760-8 (lib. bdg. : alk. paper) (cloth)
  ISBN 0-674-65761-6 (paper)
  1. Microbiology—France—History—19th century. 2. Microbiology—
Social aspects—France. 3. Pasteur, Louis, 1822–1895. I. Latour,
Bruno. Irréductions. English. 1988. II. Title.
QR22.F8L3813 1988
306'.45'0944—dc19   88-2670
                    CIP

*To Michel Serres and to all
of those who are crossing
his Northwest Passage*

Acknowledgments

The first part of this book, "War and Peace of Microbes," was translated by Alan Sheridan, and the second part, "Irreductions," was translated by John Law. The English versions were then revised and expanded by me. I thank John Law for his patience and the Ecole National Supérieure des Mines of Paris for its support.

B.L.

Contents

Part One

War and Peace
of Microbes

Introduction
Materials and Methods

Russia vs. france

From War Machines to War and Peace

On October 6, 1812, Kutuzov, general of the Russian troops, won a major battle in Tarutino over the *Grande Armée* led by Napoleon. At least, this was the impression gathered in Saint Petersburg by the czar, who offered Kutuzov a diamond star, his chief of staff, Benningsen, diamonds and a hundred thousand roubles in cash, and promotion to many of his officers. It was also the impression gathered by the French, who took this brief encounter with the Cossacks of Orlov-Denissov as a major defeat. Tolstoy, who writes about the battle in *War and Peace,* is not quite sure that it took place at all. He is sure, however, that Kutuzov did not want to fight it; rather he tried to delay it for several weeks: "Despite all his supposed power, his intellect, his experience and his knowledge of men, Kutuzov . . . could no longer restrain the inescapable move forward, and gave the order for what he regarded as useless and mischievous—gave his assent, that is, to the accomplished facts" (p. 1175).[1]

Even after accepting the fait accompli and signing the command,

3

Kutuzov kept stopping his troops every hundred feet for three-quarters of an hour! "The dispositions as drawn by Toll were perfectly satisfactory. Just as for the battle of Austerlitz it was stated—though not in German this time—that 'the first column will proceed this way and that way, the second column will proceed to this place and that place,' and so on. . . . Everything had been admirably thought out, as dispositions always are, and as is always the case not a single column reached its objective at the appointed time" (p. 1176). "That's how things always are with us—the cart before the horse!" (p. 1183). Indeed, no one during the battle knew for sure which was the horse and which the cart, the action continually drifting away from what was intended. On October 2, after Kutuzov had been forced to act against his better judgment, his signed order kept being diverted. The young officer who held it got lost and could not find the generals; eventually he arrived late at night at a mansion between the front lines where, to his surprise, the high staff were carousing. When in the morning Kutuzov got up to fight a battle he did not want to fight, he discovered to his fury that not a single soldier was prepared. No officer had received any marching orders. On the whole, however, Tolstoy considered that the battle—though not planned, not decided upon, and not fought—was a success from the Russians' point of view: "It would be difficult and even impossible to imagine any issue of that battle more opportune than its actual outcome. With a minimum of effort and at the cost of trifling losses, despite almost unexampled muddle the most important results of the whole campaign were obtained" (p. 1184).

What is this talk about attribution of responsibility, multitude of people, and missing orders? Are we not talking about strategy—the epitome of planned action—and about military chains of command—the most ordered system of direction there is? Indeed we are, but Tolstoy has forever subverted the notion of leader, strategy, and chain of command: "If in the accounts given us by historians, especially French historians, we find their wars and battles conforming to previously prescribed plans, the only conclusion to be drawn is that their accounts are not true" (p. 1184).

So what conclusion should we draw when we hear historians, especially French historians, describe not the victory or defeat of Napoleon but the victories of Pasteur, that other French genius, over the microbes? On June 2, 1881, in the little village of Pouilly-le-Fort in Beauce, Louis Pasteur defeated a terrible disease of sheep and cows,

1881

called anthrax. A friend of Pasteur's gives this account: "Pouilly-le-Fort is as famous today as any other battlefield. Monsieur Pasteur, a new Apollo, was not afraid to deliver oracles, more certain of success than that child of poetry would be. In a program laid out in advance, everything that was to happen was announced with a confidence that simply looked like audacity, for here the oracle was pronounced by science itself, that is to say, it was the expression of a long series of experiments, of which the unvarying constancy of the results proved with absolute certainty the truth of the law discovered" (Bouley: 1883, p. 439). The strategy was conceived entirely in advance; Pasteur concocted it and had every detail figured out; it went according to plan, following a strict order of command from Pasteur to the sheep by way of his assistants and the caretakers. Following Tolstoy's advice, we can say that such an account has to be false. We do not know what happened, but we can be sure that a multitude of people took part in the work and that a subtle translation, or "drift," of their intentions led them to the little village in order to watch vaccinated and unvaccinated sheep withstand tests.

We would like science to be free of war and politics. At least, we would like to make decisions other than through compromise, drift, and uncertainty. We would like to feel that somewhere, in addition to the chaotic confusion of power relations, there are rational relations. In addition to Tarutino, we would have Pouilly-le-Fort. Surrounded by violence and disputation, we would like to see clearings —whether isolated or connected— from which would emerge incontrovertible, effective actions. To this end we have created, in a single movement, politics on one side and science or technoscience on the other (Shapin and Schaffer: 1985). The Enlightenment is about extending these clearings until they cover the world.

Few people still believe in such an Enlightenment, for at least one reason.[2] Within these enlightened clearings we have seen developing the whole arsenal of argumentation, violence, and politics. Instead of diminishing, this arsenal has been vastly enlarged. Wars of science, coming on top of wars of religion, are now the rage. "Thanatocracy" is the word that Michel Serres had to forge to name our disappointment in the redeeming virtue of science.[3] Few people still believe in the advent of the Enlightenment, but nobody has yet recovered from this loss of faith. Not to believe in it is to feel that we have been thrown back into the Dark Ages.

We cannot count on epistemology to get us over this disappointment. Although epistemologies have varied over time, they have always been war machines defending science against its enemies—first in the good old days against religion, then against some of the illusions generated by too much optimism in science itself, still later against the dangers that totalitarian states represent for the autonomy of free scientific inquiry, and finally against the abuses of science distorted by politicians or corporate interests. These polemical versions of what science is and should be are convenient to fight the barbarians and keep them at arm's length; they are of no avail for describing what a polemic is and how science and war have come to be so intermingled. Epistemologists, like generals, are always one war too late. The problem is no longer to defend science against religion, abuses, brownshirts, or devious corporate interests. The problem we now face is to understand that obscure mixture of war and peace in which laboratories are only one source of science and politics among many sources. Agnosticism in matters of science is the only way to start without being trapped on one side of the many wars being fought by the guardians of science's borders.

Even if few people still believe in the naive view, courageously defended by epistemologists, that sets science apart from noise and disorder, others would still like to provide a rational version of scientific strategy, to offer clear-cut explanations of how it develops and why it works. They would like to attribute definite interests to the social groups that shape science, to endow them with explicit boundaries, and to reconstruct a strict chain of command going from macrostructures to the fine grain of science. Even if we have to give up our beliefs in science, some of us still wish to retain the hope that another science, that of society and history, might explain science. Alas, as Tolstoy shows us, we do not know how to describe war and politics any better than we know how to explain science. To offer well-conceived Machiavellian strategies to explain science is as meaningless as to write "Die erste Colonne marschiert, die zweite Colonne marschiert." Our problem in simultaneously describing wars of science, religion, and politics comes from the fact that we have no idea how to describe any war without adding to it the result of a science: strategy, history, sociology, theology, or economics.[4]

To understand simultaneously science and society, we have to describe war and peace in a different way, without ourselves waging another war or believing once again that science offers a miraculous peace of mind. Appealing to an example from an earlier period might

apriori

help us find a way out. To reestablish democracy in the troubled time of the religious wars, Spinoza had to become agnostic as far as the biblical text was concerned and to devise new ways of understanding the shocking mixture of evangelical messages and massacres. His new style of biblical exegesis in his *Tractatus Theologico-Politicus* points to a solution different from those offered by beliefs in religion or the sciences (be they natural or social).[5] Here I deal with scientific wars by using resources offered by an exegesis of scientific texts. My "Tractatus Scientifico-Politicus," instead of clearly dividing science from the rest of society, reason from force, makes no a-priori distinction among the various allies that are summoned in times of war. Recognizing the similarity among allies, I offer no a-priori definition of what is strong and what is weak. I start with the assumption that everything is involved in a relation of forces but that I have no idea at all of precisely what a force is.[6]

To make this new tack perfectly clear, I take it twice. In the first part of the book I study a series of texts taken from a famous historical battle. In the second part, I work out the principles to show how other politicoscientific mixtures can be studied in the same way. To use outdated terms, the first part of the book is more empirical, the second part more theoretical. To use more appropriate words, the first part pertains to the literary genre of sociology or social history, the second to that of philosophy. Instead of dividing the realm into those who empirically study science in the making and those who claim to guard the borders or establish the foundations of science, I combine the two, and it is together that they should stand or fall.[7]

How Are We to Dispute an Indisputable Science?

It has always seemed that if a science were not independent of politics, something would be missing and the sky would fall on our heads. To show that the sky holds up perfectly well on its own, we have to be able to prove in a particular scientific discipline that belief in the sciences, like the old belief in God, is a "superfluous hypothesis." We have to give evidence that "science" and "society" are both explained more adequately by an analysis of the relations among forces and that they become mutually inexplicable and opaque when made to stand apart.

The only way to demonstrate a proof that might win consent is to take an example that is as far removed as possible from the thesis I

Pasteur

Research + application

am trying to prove. We have to take a radical, unchallengeable scientific revolution, one that has profoundly transformed society and yet owes it very little. There are a number of reasons for believing that there is no better example than that of the revolution introduced into medicine, biology, and hygiene by the work of Louis Pasteur.

First, this revolution took place at the high point of the scientific religion. Indeed, for some decades between the Franco-Prussian War and World War One, it seemed reasonable to expect the sciences to eliminate political dispute. Second, no one—except extreme cynics—can doubt the value of Pasteur's discoveries to medicine. All of the other technological conquests have their embittered critics and malcontents—not to mention those suffering from radiation—but to prevent children from dying from terrible diseases has never been seen as anything other than an advantage—except, of course, by the microbes of those diseases. Up to our own time biology has derived its prestige from its influence on health (and most of its income from the social security system). Third, in no other scientific or technological innovation has there been so short a route between fundamental research and its rapid, far-reaching application—so much so that it is reasonable to wonder whether this is not the only example, which has been exaggerated into a general law. All the other sciences either influence only sections of society or require such a long-term mediation that in the end industry or the military always intervenes. Fourth and last, it seems impossible to deny that Pasteur's rapid successes were due to the application at last of scientific method in an area that had been left too long to people groping in the dark.[8] Most people would agree that, with Pasteur, the medical art became a science. The Pasteur blitzkrieg, in striking contrast to the physicians' and surgeons' blind struggle against an invisible enemy, reveals a convincing scientific manner, free of compromise, tinkering, and controversy. In sum, it is an indisputable case and and therefore a perfect example for my argument.

But what does "explaining" this example mean? To explain does not mean to confine the analysis to the "influences" exerted "on" Pasteur or to the "social conditions" that "accelerated" or "slowed down" his successes. To do so would once again be to filter the content of a science, keeping only its social "environment." Just as we cannot explain a myth, a ritual, or a custom connected with hunting simply by recopying or repeating it, so we cannot explain a science by paraphrasing its results. In other words, to explain the science of the

Pasteurians, we must describe it without resorting to any of the terms of the tribe.[9]

But where can we find the concepts, the words, the tools that will make our explanation independent of the science under study? I must admit that there is no established stock of such concepts, especially not in the so-called human sciences, particularly sociology. Invented at the same period and by the same people as scientism, sociology is powerless to understand the skills from which it has so long been separated. Of the sociology of the sciences I can therefore say, "Protect me from my friends; I shall deal with my enemies," for if we set out to explain the sciences, it may well be that the *social* sciences will suffer first. What we have to do is not to explain bacteriology in sociological terms but to make those two logoi once more unrecognizable.

In order to make my case, I seem to be putting myself in an indefensible position. I shall try to explain the least controversial episode in the history of the sciences without bypassing its technical content and without refusing the help that the social sciences might like to offer. The conditions of failure, at least, are clear enough. I shall fail in three cases: if this analysis becomes a sociologizing reduction of a science to its "social conditions," if it offers a satisfactory analysis of the applications of Pasteurism but not of some of its technical content, or if it has recourse to notions and terms belonging to the folklore of the people studied (terms such as "proof," "efficacy," "demonstration," "reality," and "revolution").[10]

A Method for Composing Our World

What will we talk about? Which actors will we begin with? What intentions and what interest will we attribute to them?[11] The method I use does not require us to decide in advance on a list of actors and possible actions. If we open the scientific literature of the time, we find stories that define for us who are the main actors, what happens to them, what trials they undergo. We do not have to decide for ourselves what makes up our world, who are the agents "really" acting in it, or what is the quality of the proofs they impose upon one another. Nor do we have to know in advance what is important and what is negligible and what causes shifts in the battle we observe around us. Semiotic studies of the texts of the time will do the job of *interdefinition* for us. Take, for instance, this article by Tyndall: "Consider

Enquiry for Composing World by Brano Latour

all the ills that these floating particles have inflicted on mankind, in historic and prehistoric times . . . This destructive action is continuing today and continued for centuries, without the slightest suspicion as to its causes being permitted to the sick world. We have been struck by invisible scourges, we have fallen into ambushes, and it is only today that the light of science is reaching those terrible oppressors" (1877, p. 800).[12]

Without any other presuppositions, we can take this sentence as our beginning and study it with the tools offered by semiotics. Tyndall defines actors. Are they human or nonhuman? Nonhuman. What do they want? Evil. What do they do? They lie in wait. Since when? Since the beginning of time. What has happened? An event: they have become visible. What has made them visible? Science, another actor, which must in turn be recorded and defined by its performances.[13]

The fact that we do not know in advance what the world is made up of is not a reason for refusing to make a start, because *other* storytellers seem to know and are constantly defining the actors that surround them—what they want, what causes them, and the ways in which they can be weakened or linked together. These storytellers attribute causes, date events, endow entities with qualities, classify actors. The analyst does not need to know more than they; he has only to begin at any point, by recording what each actor says of the others. He should not try to be reasonable and to impose some predetermined sociology on the sometimes bizarre interdefinition offered by the writers studied. The only task of the analyst is to follow the transformations that the actors convened in the stories are undergoing. For instance, an anonymous editorial, written just after the Franco-Prussian War, states: "It is science and the scientific spirit that have conquered us. Without a complete resurrection of the great French science of former times, there is no possible salvation" (1872, p. 102).

Is this an "ideological" rendering of what really caused the French defeat?[14] Is it a "false" representation of what happened? Is this a pure "expression" of late nineteenth-century scientism? The analyst does not have to know. In 1872 the editorialist attributed defeat to a lack of science. This attribution is enough for us to be able to follow the drift at work in the editorial. You want revenge? asks the writer. For that, you need soldiers. In order to have soldiers, you need healthy Frenchmen. But what is it that watches over health? Medicine. And what does medicine itself depend on? The sciences. And what are the sciences in turn made up of? Money. And where does money come from? The state budget. But parliamentarians are right now discussing subsidies for research: "The cuts spare those who shout the loudest," writes the editorialist. Hence his advice: write to your deputies, so that the government will not cut the budget, so that there will be laboratories, so that there will be sciences, so that there will be medicine, so that . . . so that . . . and so that we can wreak our revenge at last. We do not have to know what this writer "really" wants any more than we have to know what the generals surrounding Napoleon or Kutuzov "really" wrote in their marching orders.[15] It is enough that the writer has made up his editorial in such a way that a reader who wants revenge with all his heart can now be requested to petition his deputy against budget cuts. This movement of *translation* is enough for us.[16] We had our eyes fixed on the "blue line of the Vosges." Now we have them riveted on the sheet of paper arguing that Alsace will be won back *more quickly* by means of this passage through the scientific budget.

The method I use here consists simply in following all these translations, drifts, and diversions as they are made by the writers of the period. Despite my search for complication, I could find no more than this simple method. Semiotics provides me with a discipline, but since it is too meticulous to cover a period of fifty years and thousands of pages, the semiotic method is here limited to the interdefinition of actors and to the chains of translations.[17]

I apply these simple tools to the analysis of three periodicals: the *Revue Scientifique, Annales de l'Institut Pasteur,* and *Concours Médical.* The *Revue Scientifique,* a general weekly review founded in the mid-nineteenth century and written by scientists themselves for a wider educated public, falls somewhere between *Scientific American* and the general-interest pages of *Science.* I read through the whole of the journal from 1870, the year of France's defeat, to 1919, the date of the revenge but also of a terrible defeat at the hands of influenza. I did not confine myself to a particular science but recorded all the references made by the authors to diseases, biology, health, Pasteur, microbes, doctors, and hygiene. For each of the relevant articles I sketched the interdefinition of the actors and the translation chains, without trying to define a priori how the actors were made up and ranked. Without being exhaustive, I nevertheless recorded the great majority of the allusions, however distant, to Pasteur and his microbes throughout the pages of the *Revue.*

The *Annales de l'Institut Pasteur,* founded in 1887, is the official scientific journal of the Institut Pasteur. In this case all articles from 1887 to 1919 were treated and codified according to a single specification that was also borrowed from semiotics. Unlike the study of the *Revue,* this one involved reading the complete corpus and thus offered a precise idea of the official scientific output of the Institut. The *Concours Médical,* a periodical published by a French medical union, was studied only for the crucial years 1885–1905. In this case I recorded only the explicit allusions to Pasteurism, without trying, as in the *Revue,* to retrace the path of implicit translations as well.

Since the documentary material is limited to these three journals, my effort to explain bacteriology and French society simultaneously may be judged solely on this basis. Pasteur said he could not claim the honor of being a surgeon; I cannot claim that of being a historian. This undertaking does not purport to add anything to the history of science, still less to the history of the nineteenth century. I use history as a brain scientist uses a rat, cutting through it in order to follow the mechanisms that may allow me to understand at once the *content* of a science and its *context.* For this reason the presentation of the documentary materials does not follow the historical path but rather the network of associations that slowly make up the Pasteurian world. Fortunately, the period offers us a great many control groups that react differently to Pasteur's enterprise. Hygienists, biologists, surgeons, sanitary engineers, veterinary surgeons, physiologists like Claude Bernard, medical doctors, and military doctors, as well as tuberculosis, cholera, diphtheria, tetanus, yellow fever, rabies, and the plague, all move according to different paths, offering us the sort of interesting confusion that Tolstoy describes in the battles of his book. Here I contrast the different control groups with one another, so that each argument about context or content can be replaced by a new linkage between society and its sciences.

Looking
for
network
association

Chapter 1

Strong Microbes and
Weak
Hygienists

Is It Necessary to Speak of "Pasteur" or Even of Pasteur?

The counter-example that I have chosen to study is so obviously incontrovertible because of the way it is habitually formulated: "the revolution introduced into medicine by Pasteur." What we have here is an attribution of cause and time. We might also say that it represents a dominant point of view—a point of view that was therefore victorious in a battle fought with other agents pursuing other aims at other times. Do we have to speak of Pasteur when we speak of hygiene and medicine in the late nineteenth century? It is not immediately obvious that we do. Pasteur's position in this is rather like Napoleon's in that treatise on political philosophy which Tolstoy wrote under the name of *War and Peace.*

In that book, Tolstoy summons up hundreds of characters to give depth to what for him is the essential question: What can one man do? What does a great man like Napoleon or Kutuzov really do? It takes Tolstoy some eight hundred pages to give back to the multitude

13

the effectiveness that the historians of his century placed in the virtue or genius of a few men. Tolstoy succeeded, and the whole of recent history supports his theories as to the relative importance of great men in relation to the overall movements that are represented or appropriated by a few eponymous figures. This is true at least where *politicians* are concerned. When we are dealing with *scientists*, we still admire the great genius and virtue of one man and too rarely suspect the importance of the forces that made him great. We may admit that in the technological or scientific fields a multitude of people is necessary to *diffuse* the discoveries made and the machines invented. But surely not to create them! The great man is alone in his laboratory, alone with his concepts, and he revolutionizes the society around him by the power of his mind alone. Why is it so difficult to gain acceptance, in the case of the great men of science, for what is taken as self-evident in the case of great statesmen?

If Tolstoy is indignant against the Napoleonic hagiography, what are we to say of the attributes that the French have given to Louis Pasteur from the outset? He did everything; he regenerated, revolutionized, created the new medicine, the new biology, the new hygiene. Landouzy writes: "Never will a century have worked for the century whose dawn you will soon be welcoming, as did the century of Pasteur" (1885, p. 107). It is not given to everybody to become a century, any more than it is to have one's name on the principal street of every town and village in France, or to prevent people from spitting, to persuade them to dig drains, to get vaccinated, or to create serotherapy. Pasteur did everything, by his own power, or at least through the power of his ideas. Such a view is no more tenable than is the statement that Kutuzov defeated Napoleon. Of all great men, it must be said: "The only conception capable of explaining the movement of peoples is that of a force commensurate with the whole movement of the peoples. Yet to supply this conception various historians assume forces of entirely different kinds, all of which are incommensurate with the movement observed."[1] The notion of power, be it that of ideas or of political clout, is one of these misconceptions: "So long as histories are written of separate individuals, whether Caesars, Alexanders, Luthers or Voltaires, and not the histories of *all*—absolutely *all*—those who take part in an event, it is impossible not to ascribe to individual men a force which can compel other men to direct their activity towards a certain end. And the only conception of such a kind known to historians is the idea of power" (Tolstoy: 1879, p. 1409).

If the whole of Europe transformed its conditions of existence at the end of the last century, we should not attribute the efficacy of this extraordinary leap forward to the great genius of a single man. Yet we can understand how he followed that movement, accompanied it, sometimes preceded it, and then was offered sole responsibility for it (at least in France). Pasteur's contemporaries, the Pasteurians, and French historians were not unaware of this problem. They all admitted that Pasteur did not do everything "alone," but they quickly went back on this admission by supposing that Pasteur did everything "potentially" or that the rest was to be found in Pasteur's ideas *in potentia*. "There was a man," says Bouley, one of Pasteur's hagiographers, "and to tell of the great things that I have to relate, I shall borrow one of Bossuet's famous periods, there was a man of incredible depth of mind" (1881, p. 546). Indeed, we are tempted to fall down on our knees in admiration, since the rapid, complete transformation of a society is attributed to the "thought" of one man. "Are you not confounded," Trélat exclaims, "by the force of the genius who could win such battles?" (1895, p. 170). Yes, of course we are confounded— if we confuse the force of a man with that *attributed* to him; if we confuse Pasteur with "Pasteur," whom from now on I will place between quotation marks so as not to confuse him with his homonym. Why should we still do for Pasteur's genius what we no longer do for Napoleon's or Rothschild's? If we find it easy enough to deal with the Russian campaign in terms of sociology or economics, why are we so reluctant to apply sociology to Pasteurian bacteriology?

The reason for this hesitation is simple. The French analysts do not even hesitate. For them there is nothing to analyze. Indeed, they almost invariably suppose that where science is concerned, the diffusion of an idea, a gesture, a technique, poses no particular problem; only the constitution of the idea or gesture is problematic. Their idea of society is borrowed from classical mechanics: techniques, endowed with inertia, a resistance to force, always retain that property, and can only lose it in the course of successive shocks. With such a model, we must attribute to Pasteur's laboratory the *totality* of the force and regard as inert masses all the social groups that are capable only of transmitting the force or absorbing part of it (it is said of them that they "adapt to progress" or "resist"). But it must be clearly understood that in social physics there is no law of inertia. To convince someone that an experiment has succeeded, that a technique is effective, that a proof is truly decisive, there must be *more than one actor*. An idea or a practice cannot move from A to B solely by the force that A gives it; B must seize it and *move* it. If, to explain the "diffusion" of Pasteur's ideas, we had nothing more than the force of Pasteur and his collaborators, those ideas would never have left the walls of the Ecole Normale laboratory and would not even have *entered* them. An idea, even an idea of genius, even an idea that is to save millions of people, never moves of its own accord. It requires a force to fetch it, seize upon it for its own motives, move it, and often transform it.[2]

This vision of things poses no particular problem, except that it regards all the sites where a particular practice is diffused as made of autonomous agents rather than of inert masses passively transmitting a force. Tolstoy has to reconstitute Russian society and the autonomy of all his characters in order to take from Caesar the things that are not Caesar's. Similarly, freedom of action must be given back to all the agents of French society in order to *decompose* Pasteur's efficacy. Here lies the problem: to make a *sociology* of bacteriology, one needs a *society*.[3]

When I began to read the *Revue Scientifique* after the defeat of 1870, I was surprised to observe that little is said of Pasteur and even less of his ideas. He is not yet the intercessor that he would later become. His name is not yet associated with anything relating to disease. Other things are discussed, and the evidence presented does not come from his laboratory.

The Indisputable Conflict between Health and Wealth

Although the authors of the *Revue* rarely speak of Pasteur or discuss his ideas, they view one idea as so indisputable that it becomes the premise for all the arguments to be found in the *Revue* from the first number of the new series, begun just after the siege of Paris, to the last number studied, in December 1919. This idea, which they regard as overwhelmingly, universally self-evident, is "the urgent need for regeneration."

"It is to the doctors that a large part of the work of regeneration falls if such work can ever be carried out, for the first condition of force is the number and vigor of the citizens," writes Algave, director of the *Revue* (1872, p. 102). From July 1871 on, Pasteur claims, he was mobilizing science for the cure of "the Prussian canker" (1871, pp. 73–77). It was not only France, humiliated and defeated, that had to be regenerated; it was also mankind in general and, more particularly, the urban masses. In 1872 Stokes sums up the state of the new British medicine, already highly developed, and defines the new deal of political action: "Instead of arguing over principles and seeking the absolute, this people [the British], gifted with a great practical sense, is painstakingly erecting the props that support the old social edifice and is making it inhabitable for the new populations" (1872, p. 14).

We could not have a better definition of the program of reforms—of "social" not "political" reforms, the author insists on pointing out—in which first public medicine and then the biological sciences that will enable it to advance play an important role. Stokes continues: "What an opportune moment to apply all those scientific forces to preventative medicine and consequently in the social order! There are hundreds of millions of subjects of the Crown of England, whose domestic habits seem to be scarcely above those of the lower animals, and an enormous field of misery, physical and moral degradation, and a constant source of destruction that may extend to the confines of the earth and return back against the West, where the noblest race of men is to be found" (1872, p. 20).

Many historians have insisted on this obsession of the time with the regeneration of man.[4] It serves as a premise for all the articles in the *Revue* not only on medicine but also, over the years, on gymnastics, colonization, international trade, education, the economy, warfare, and above all, the depopulation of France, which Richet calls "the greatest peril that the French nation has ever had to face, at any period in its history." All the articles repeat the view in one way or another that what we need are strong men: "The first concern of statesmen today is the reconstitution, the reorganization of human life. The independence, the very existence of the country in the near future, is at stake" (Decaisne: 1875, p. 933). It should be stressed that all these quotations are taken from authors who are extremely dubious about contagionist theories, have hardly heard of asepsia, and are writing some fifteen years *before* the slightest application of bacteriology to human medicine.[5]

But what is this movement itself based on? This question, raised by historians, does not have to be answered by the semiotic method I have chosen to follow. Since all the writers take this basic link between health and wealth as settled, since all of them take hygiene as the "addressee" of all their articles, and since this character was constituted before the period under consideration, I could move on and begin with the irruption of the microbe.[6] Fortunately, it is possible

to sketch the backdrop against which the whole Pasteurian drama-
turgy unfolds. This sketch can be made in two successive stages: the
first presents an infrastructure that explains the energy accumulated
during the period; but the second highlights another science, another
group of scientists who have already prepared the ground for the
arrival of the Pasteurians.

For those who cannot accept any story unless it has an "infrastructure," it is possible to give the "cause" of the whole Pasteurian
adventure. In simple terms, Frazer sums up this motive force of the
period, the "primum movens" that unleashed all those energies but
was itself moved by nothing and discussed by nobody.[7] The conflict
between health and wealth reached such a breaking point in the mid-
century that wealth was threatened by bad health. "The consumption
of human life as a combustible for the production of wealth" led first
in the English cities, then in the continental ones, to a veritable "energy
crisis." The men, as everyone said constantly, were of poor quality.
It could not go on like that. The cities could not go on being death
chambers and cesspools, the poor being wretched, ignorant, bug-
ridden, contagious vagabonds. The revival and extension of exploitation (or prosperity, if you prefer) required a better-educated population and clean, airy, rebuilt cities, with drains, fountains, schools,
parks, gymnasiums, dispensaries, day nurseries. By the time that concerns us, none of this was controversial. It was the starting point from
which hygienists set out to discover latent forces and to set up particular strategies.

The concepts of infrastructure thus regards immense energies as
being mobilized by this contradiction of health and wealth throughout
Europe. Such an upheaval of cities was seen not as a revolution but
as a harmonization, in Stokes's words, between "national health" and
"national prosperity and morality" (1872, p. 20). The favorite metaphor of the time, the difference in potential, defined a vast energy
source into which all the actors of the period could plug themselves
in order to advance their concerns for the next fifty years. This immense reservoir of energy was a force of the kind demanded by Tolstoy, one that was commensurate to the social body itself. In this
infrastructure story, Pasteurians are one of the many groups that use
the same difference in potential, even if the word "Pasteur" came to
designate in France the whole of this universal movement of regeneration.

Every time historians speak of an infrastructure that can explain

Historians + Sociologists of Science

the development of a science, the sociologist of science, devious and suspicious, looks for what *former* scientific professions have already done to create this vast reservoir of energy. Often no study is available, and the sociologist has to abandon the ground and believe, like anyone else, in the idea of a preexistent social context, at least for the period and the science he is *not* studying. Fortunately, Coleman has made an excellent study of the period just preceding mine.[8] In this study we see another group of scientists, another profession, led not by Pasteur but by Villermé, busy creating this famous "infrastructure" and this famous conflict between Health and Wealth.[9] Before the period under consideration we do not have a "longue durée" that would act as a cause to push or pull the Pasteurians, but we have Villermé and his friends constituting, through the new profession of scientific hygiene and through the elaboration of national statistics, a link between mortality and degree of wealth. This link had also to be created, like the future link between laboratory and medicine or between attenuated microbes and diseases.[10] Without the creation of statistical bureaus and "tableaux," without the application of political economy to this sociomedical problem, the "difference in potential" would not have existed. The social context of a science is rarely made up of a context; it is most of the time made up of a *previous* science.[11]

Hygienists, the Disputed Interpreters of Regeneration

I shall say no more about this "infrastructure," since it inspired the articles without itself ever being discussed. Yet I must say something about the first translators of this great conflict between health and wealth, the hygienists. Actually, the *Revue* does not define who they are. It speaks of hygiene, the "sender," as the semioticians say, of all the actions on health. The boundaries of hygiene are vague, and this vagueness is precisely what allows its practitioners to express more or less everyone's interests and, very soon, those of the Pasteurians. Here again we must not, in our study of the texts, be more precise than the *Revue* itself. For our purposes, hygienists are all those who call themselves hygienists.

Hygiene in the *Revue Scientifique* can be defined as a *style*. An article, especially a scientific one, is a little machine for displacing interests, beliefs, and aligning them in such a way as to point the reader, almost inevitably, in a particular direction. Scientific rhetoric often channels the reader's attention in a single central direction, like a valley cutting through mountains. But the rhetoric of the hygienists does not possess this great flow. It has no central argument. It is made up of an accumulation of advice, precautions, recipes, opinions, statistics, remedies, regulations, anecdotes, case studies. It is, indeed, an accumulation. A hygienist like Bouchardat always adds, without subtracting anything at all. The reason for this style, which in the literary criticism of an earlier day would be called "involved" or "cautious," is simple. Illness, as defined by the hygienists, can be caused by almost anything. Typhus may be due to a contagion, but it may also be due to the soil, the air, overcrowding. Nothing must be ignored, nothing dismissed. Too many causes can be found side by side to allow for any definite position on the matter. Everything must be considered. "The role and variety of the causes of typhoid make it necessary to combat them by equally varied and numerous means" (Colin: 1882, p. 397). It was not out of ignorance but on the contrary out of an excess of knowledge that the hygienists accumulated their opinions. None of them is absolutely certain, they admit, but none of them can really be abandoned. Bouchardat makes the ingenuous admission: "I do not spend my hours of sleep in intensely choleraic places." He advises the use of disinfectants but adds, "they must not allow us to ignore evidence that is not understood but is based on strict and repeated observation" (1883, p. 178).

To make fun of this style would be to fail to understand the nature of an all-round combat. If anything can cause illness, nothing can be ignored; it is necessary to be able to act everywhere and on everything at once. The style reflects the action planned by the hygienists. Many of the characteristics of so-called pre-Pasteur hygiene are to be explained by this situation. The hygiene congresses were, like Bouchardat's style, an attic in which everything was kept because sometime it might come in handy. In 1876, for instance, the subjects under discussion included water, lifesavers, gymnastics, women's work, "methods of developing among the laboring classes a spirit of thrift and the saving habit," alcoholism, and working-class housing (Anon.: 1876, p. 400). These congresses were a catchall, because illness could be caused by anything and because scientists had to be ready to set off enthusiastically in any direction.

The consequences were predictable. Articles on hygiene in the *Revue* were shot through at first by an astonishing combination of hubris and discouragement. Both had the same cause. Since anything might cause illness, it was necessary to act upon everything at once, but to act everywhere is to act nowhere. Sometimes the hygienists give a definition of their science that is coextensive with reality. They claim to be acting on food, urbanism, sexuality, education, the army. Nothing that is human is alien to them. Even the human being is too narrow a field; they must concern themselves also with air, light, heat, water, and the soil (Trélat: 1890, pp. 705–711).

But to understand everything is to understand nothing. So the same articles reveal a sense of division and "abasement" (Landouzy: 1885, p. 100). Indeed, the fundamental problem of the hygienists is that this multiplicity, so short on remedies and details, did not protect them against failure. However much they might take precautions against everything and observe everywhere, disease returned, as if no fixed causes could be attributed to it. On each of its returns another cause had to be added. The surgeon Kirmission writes, after emerging from that period: "So we accumulated all the precautions of general hygiene, but failed to remove the purulent infection from the wards . . . In demonstrating the inanity of all the discussions on hospital hygiene as a way of preventing hospital infection, experience necessarily cast profound discredit on the pious wishes of the surgeons" (1888, p. 296).

For all these reasons it was necessary to speak of "morbid spontaneity." This doctrine, which is ridiculed today, corresponded perfectly to the style, mode of action, and facts, since disease appeared sometimes here, sometimes there; sometimes at one season, sometimes at another; sometimes responding to a remedy, sometimes spreading, only to disappear as suddenly. This strange, erratic behavior was well recorded by statistics, the major science of the mid-nineteenth century, which corresponded perfectly to the analysis of such impalpable phenomena.[12]

In view of these problems, it was also logical that any article on contagion, on the microbe as "external cause" of disease, on the law that "a microbe equals disease," should appear so derisory. To any argument on contagion itself a budding hygienist could always oppose a hundred counterexamples. This disproportion between the problems of the hygienist and the simplistic character of the doctrines of contagion helps to explain how the Pasteurians had to transform the microbe in order to convince the hygienists. The hygienists formed the vanguard of a huge, century-old movement which had already transformed the British system of health and which claimed to be spreading everywhere in order to act on all the causes of ill health.

But by its very scope and ambition this movement remained weak, like an army trying to defend a long frontier by spreading its forces thin.

There was no way of concentrating the movement's forces at a few points only. It could not *ignore* the details that it had accumulated for hundreds of years, unless it could hierarchize them in order of importance. As soon as hygiene became modern, that is, turned the hygiene that had preceded it into "ancient" hygiene, it was by its very "lightness" that it was recognized. As Bouchardat remarks: "If, at the beginning of this century, we strove to understand everything in hygiene, today we must leave to one side a mass of useless or unprovable details" (Landouzy: 1885, p. 100).

Was it possible to define in advance, negatively, that excess of force which, retrospectively, hygiene seemed to lack? I think so. What was needed was a source of forces to explain the astonishing variability of morbidity, its spontaneity, and its local character. In order to interest the social movement of which the hygienists were the spokesmen, a doctrine was needed that explained the *variation* of the virulence in terms compatible with the problems involved in transforming the towns and the living environment to which the hygienists had devoted their attention. This was not simply an "intellectual" requirement. In the absence of such a focal point, all the energies of the social movement translated by the hygienists were dissipated through thin networks, all of them relatively equal in size and therefore doomed to extinction before being able to reach any of the great goals that the movement had set itself. At the time—that is, before Pasteur had made himself necessary to the hygienists—one thing was certain: the doctrine of contagiousness was inadequate to fulfill the hygienists' goals.[13]

The Movement of Hygiene Left to Itself

To speak of hygiene was already to take up a position. It was to go back. It was to try to distinguish retrospectively what had been intentionally confused. To try to see what the hygienists would have been before they became closely involved in Pasteurism was, as it were, to set a pyramid that had been standing on its point back on its base. Tolstoy was right here, too. A crowd may move a mountain; a single man cannot. If, therefore, we say of a man that he has moved a mountain, it is because he has been credited with (or has appropriated) the work of the crowd that he claimed to command but that he also followed. An enormous social movement ran through the social body in order to reconstruct leviathan in such a way that it could provide shelter for the new urban masses.[14] The hygienists used this movement to attack disease on every side or, in their language, to act "on the pathogenic *terrain*." The Pasteurians, who numbered, let us not forget, no more than a few dozen men at first, set out in turn to direct and to translate the hygienist movement. In France, the result was such that the hygienist movement came to be identified with the man Pasteur, and ultimately, following a very French habit, the man Pasteur was reduced to the ideas of Pasteur, and his ideas to their "theoretical foundations." In the end, then, what emerged was that inverted world stigmatized by Tolstoy: a man moves a mountain by his genius alone.

The first people to undergo this reversal were the readers of the *Revue Scientifique*. Indeed, it is almost impossible to discern a "pure" hygienist movement completely separate from the expression given it by the Pasteurians. However, even at the cost of a fiction, it is crucial to rediscover, at least in imagination, the crowds moving the mountain, so that we can understand later how the Pasteurians came to be their spokesmen and were regarded as the "cause" of the movement. Where would the hygienist movement have gone without Pasteur and his followers? In its own direction. Without the microbe, without vaccine, even without the doctrine of contagion or the variation in virulence, everything that was done could have been done: cleaning up the towns; digging drains; demanding running water, light, air, and heat.[15] Pettenkoffer, who swallowed cholera bacilli without becoming ill but made Munich a healthy city through large-scale public works, is for everyone the eponym of this attitude in history. Verne's *Les 100 millions de la Bégum,* which contrasts Hygié, the healthy French town, with Noson, the unhealthy "Boche" town, without the slightest mention of a microbe, is the literary counterpart of Pettenkoffer. The fulcrum provided by bacteriology should not let us forget that the enormous social movement was working for that mixture of urbanism, consumer protection, ecology (as we would say nowadays), defense of the environment, and moralization summed up by the word *hygiene*. If we do not restore the power ratio between the social movement at work throughout Europe and the few bacteriological laboratories, we cannot understand the real contribution of those laboratories, just as we cannot understand what Kutuzov did if we attribute to him the entire movement of his army.

Nowhere is the disproportion between that hygienist movement and the "small group" of Pasteurians more clearly seen than in an article of 1884 on the hygiene exhibition in London. Such exhibitions, which were frequent at the time, "bring together," reports the journalist, "several fairly complex orders of knowledge, constituting in short whatever may render life healthy and even comfortable" (Anon.: 1884, p. 386). There were tastings of Liebig soups (German chemistry), refrigerated meat (British thermodynamics), and pasteurized milk (French microbiology). People admired hygienic clothes, orthopedic shoes, light-colored furniture that could be dusted easily, filters to purify water, bidets to wash one's behind, and flushing systems to evacuate excrement. Plans were discussed for drainage, ventilators, windows, heating apparatuses—anything that would allow the four elements to circulate freely. There were life-sized models of hygienic—that is, airy and clean—houses, hospitals, ambulances, stretchers, crematoriums, classrooms, and even desks.

Bacteriology was indeed present in the exhibition, in an interesting way. To begin with, it was dispersed throughout several sections: the Chamberland filter, from Pasteur's laboratory, was placed in the series of filters proposed by industrialists; pasteurized milk was part of the new milk circuit; the incubator, deriving from "the experiments of Koch, Wolflugel, and Pettenkoffer in Germany, and Vallin in France," had been developed by industrialists and was part of the legal disinfection departments, each of which had its own stand. Disinfectants also had their place: "The current cholera epidemic has given new vigor to the study of disinfectants, a study that so far has given far from satisfactory results and in which we will now have to take greater account of the physiological and morbid properties of the specific organisms of contagious diseases" (p. 394). "To take greater account of"—that says everything. The products of bacteriology were added to hygiene like some spice that increased its local effectiveness.

But this science was present in another way, too: "In the middle of the main room are found the objects sent by M. Pasteur, by the laboratory at Montsouris [run by Miquel, a microbiologist], and by the municipal chemistry laboratory of the city of Paris." The author, of course, tries to reduce the whole of the exhibition to this section, because he is a scientist and a nationalist. This laboratory, he writes, "has made many people say what has been said aloud by an American: 'There was more hygiene in the French section than in the rest of the exhibition put together'" (p. 397). This patriotism and bacteriocentrism are honorable enough, but they contradict the whole of the article. Pasteur's laboratory was only one among many others, and it was surrounded by the exhibits of innumerable industrialists, reformers, leagues for the propagation of this or that, professions, and skills. It could not be reduced to that proliferation of exhibits, but neither could the entire exhibition be reduced to the laboratory.

To reconstruct Pasteurism, it has to be said, even with a certain degree of exaggeration, that what the hygienist movement did with Pasteur it would have done anyway without him. It would have made the environment healthier. The vague words "contagion," "miasma," and even "dirt" were enough to put Europe in a state of siege, and it defended itself by cordons sanitaires against the infectious diseases. Of course, terrible diseases got through the cordons, but sometimes there were victories, and that was no small achievement. This way of isolating hygiene and trying to discover where it was going on its own was not so arbitrary, since, after all, there is still a good deal of controversy about the causes of the remarkable improvement in the health of Europeans between 1871 and 1940.[16] The improvement is still being attributed to new causes and new agents whenever a new group sets out to weaken the position of medicine or the role of science in medicine or to redistribute in a different way the respective roles of therapy and prevention. The general rise in the standards of living and nutrition, combined with "elementary" hygiene, would be enough for some to explain most of the astonishing therapeutic successes that the Pasteurians had attributed to the science founded by Pasteur.

Even if this conflict does not concern us here, one thing is clear. It is the hygienist movement that defined what was at stake, prescribed the aims, posed the problems, demanded that others should solve them, distributed praise or blame, and laid down priorities. It is also the hygienist movement that galvanized people's energies, found the money, and offered those who served it troops, goals, problems, and energy. This is a crucial point, for it allows us to extract from the magic circle of "science" much of what we rather hastily call "its contents." The subjects that are studied and the problems that are given priority make up, as we know, most of a discipline. The Pasteurians were to arrive on the scene like players in a game of Scrabble. The "triple" words and "double" words were already marked and laid down. The Pasteurians translated these stakes and rules into their own terms, but without the hygienists, it is clear that very little would have been heard about them. The Pasteurians would have done something else.

If that penultimate sentence seems dubious, we have only to read the British or American histories of the period.[17] Bacteriology, common in these works, is far from being the source and cause of hygiene; it is merely a ripple on the surface, an aspect, doubtless a crucial support of social hygiene, but no more than that. In these histories Pasteur himself is merely one bacteriologist among others, and they emphasize not so much Pasteur's ideas as certain practical applications considered by the authors to be particularly important, such as methods of culture, incubation, and inoculation.

The Hygienists Believed Pasteur without Question

The *Revue Scientifique* reveals first of all the size of the social movement for regeneration, indicates the translator of this movement, hygiene, and shows how uncertain and controversial the hygienists were. It also shows, but less clearly, the disproportion that existed between the hygienists and the Pasteurians. Finally, study of the *Revue* explains why it is so difficult to decide how much should be attributed to each group, or even to avoid the impression of a revolution.

If we recall the way in which different authors place Pasteur when they begin to talk about him in the early 1880s, we are struck by one overwhelming fact: they do not argue over him; they trust him entirely. We may of course attribute this trust to the quality of the evidence produced by Pasteur, to the efficacy of the treatments proposed—in short, to the truth of Pasteurian science. But this is quite impossible, first because, when others were presented with the same evidence, it was regarded as disputable and second because the trust accorded to Pasteur was so great that it must have been based on something else.[18] If we convince someone of something, we must share the efficacy of that conviction with the person whom we have convinced. But if someone catches on at once, takes over what we have said, and immediately generalizes it, expands it, and applies it to other things than those we originally had in mind, then we must attribute a *greater* efficacy to the person who has understood than to the one who has been understood. For Pasteur's arguments in the *Revue Scientifique* were not exposed to sarcasm and doubt; they were seized on avidly and extrapolated well beyond the few results that he himself was defending. The avidity of those who seized on what he said gives us some idea of the extent of the social movement whose main outlines I have been tracing. Let us look at it more closely.

In 1871 Chauveau writes in the *Revue* on the contagious diseases: "We are already pressing forward, overtaking one another on the road that leads to the most useful conquests of modern science" (1871, p. 362). In 1876, well before the first studies on rabies, Tyndall considers that the revolution carried out by Pasteur is *already complete*. "It is only a question of time." His confidence is such that he looks to the future "with the interest of a man who sees a principle spreading and becoming established that is destined to deliver medicine from the reproach of empiricism, to raise it to the rank of a true science, and to deliver to the doctors those invisible enemies, as the celebrated Cohn called them, who hide in the air we breathe and the water we drink." He adds: "I doubt whether in ten years from now there will remain in England a single doctor willing to support the ideas that they thought fit to advance against Pasteur [in denying contagion]" (1876, p. 560). The year 1886 was not a bad prediction. But Tyndall did not have to be a prophet to propose such a date. It was a matter of elementary technological forecasting on the basis of a research program that had already been initiated; all that had to be done was to wait and pick the fruit.

The British were of course more advanced than the French, but Pasteur's compatriots were not lagging behind. Even the prudent Bouchardat did not hesitate to write on the subject of the plague that it would be necessary "to isolate and cultivate the microorganism as Pasteur would have done" (1879, p. 918). Richet, editor of the *Revue* and a convinced Pasteurian, supported in 1880 the project for a national award for Pasteur, "so that Monsieur Pasteur may give to his researches into the contagious diseases of animals all the developments it potentially has" (1880, p. 35).

This was written in 1880. How could Richet know how many developments the few laboratory cases would have? If someone bet a token and someone else immediately bet a hundred, how are we to understand the confidence of the second bettor? The prodigious developments given by Richet and his peers to what Pasteur was proposing must be attributed to *them*. They knew that they were going to amplify these propositions with their own. After Pouilly-le-Fort, Richet extends the efficacy of the vaccine without the shadow of a doubt: "Anthrax will soon be a thing of the past" (1881, p. 161). After the cure of Joseph Meister alone, Richet exclaims: "And now that we can cure rabies, we have only to expand and facilitate the treatment" (1886, p. 289). The year before, Landouzy exclaims: "Yes, gentlemen, the day will come when, thanks to militant, scientific hygiene, diseases will disappear as certain antediluvian animal species have disappeared (1885, p. 107).

Yet no disease had disappeared. The confidence in the "way laid down" by Pasteur must therefore derive from something other than the facts, hard facts. The confidence was not one that came only *from* Pasteur, but one that flowed back *on* to Pasteur and which he made full use of. The Pasteur of the *Revue Scientifique* was not an obscure hero who was fighting alone against all and who had to convince his irremediably skeptical adversaries step by step (Vallery-Radot: 1911). No, he had only to open his mouth, and others would turn his results into generalizations about every disease. A peculiar revolution indeed! To be sure, once Richet became its editor, the *Revue* was on Pasteur's side and defended him "beyond the limit of all scientific prudence," one might say. When a timid challenge is raised, Richet, his flank guard, writes with condescension: "It is no bad thing if a discordant voice is raised amid a concert of praise. Perhaps it will encourage M. Pasteur to provide us with a few new discoveries as fruitful as the previous ones" (1882, p. 449).

That the *Revue* and all its authors should be so partial, so chauvinistic, so imprudent, shows the extent to which trust was placed in Pasteur, exactly as money is placed in a trust fund. The reader must now understand that, if the hygienist movement had not been presented first, it would have been necessary to attribute a "prodigious efficacy" to the experiments of Pasteur himself. Generally "science" is never to be explained by itself. It is an ill-composed entity which excludes most of the elements that allow it to exist. The social movement into which Pasteur inserted himself is a large part of the efficacy attributed to Pasteur's demonstrations.

Even the Pasteurians who were most determined to spread the myth of a Pasteur struggling alone against the shades of obscurantism are forced to recognize the unanimity with which his experiments were received. Bouley, for example, writes: "Before such results [at Pouilly-le-Fort], there was no longer any room for doubt, even on the part of his most thoroughgoing opponents, who were compelled to fall silent, and the convictions acquired were immediately expressed by a sort of avidity for this new vaccine under the protection of which the farms of anthrax-ridden regions were impatient to place their flocks and herds." Bouley adds: "Justice is often slow in coming for inventors, often its progress is so limping that their lives are not long enough for them to see it done. M. Pasteur, I can now name him, has been privileged enough to see it accelerated in his case" (1881, p. 549). It should be said that in Pasteur's case justice got carried away, since it soon attributed to him what he did not in fact do and paid him the homage of the entire hygienist movement. A columnist, probably Richet, writes: "It is often difficult for contemporaries to judge the enormous progress that is being made and to have an opinion of a recent discovery that will be confirmed by posterity. However, we have, in scarcely a few months, witnessed the blossoming of a great discovery, judged as such and on the importance of which there has been unanimity" (Anon.: 1881, p. 129).

On a Few Dissenters: Koch and Peter

Nothing demonstrates better the unanimity of the crowds that followed Pasteur and seized on his results than the few people who had what might be called the courage to oppose them. Although opponents were numerous enough in the Académie de Médecine, where Pasteur sought them out with a violent rhetoric, there were only two in the *Revue*: Peter, the old-fashioned French physician, and Koch, the modern-minded German physician.[19] Although these men were entirely opposed in their beliefs, they had the same criticism of Pasteur: he generalized too hastily on the basis of a few inadequately clarified cases.

Peter has been described as an obscurantist buffoon, but he was the only one to put up any kind of a fight against Pasteur's medical coup d'état. Peter fought against the "microbic furia," against what seemed to him to be a "torrent," even "an intellectual cholera against which sanitary measures must also be taken." And "that is why," he adds, "I am for resistance." It was he who was resisting an invasion, not the Pasteurians, who were resisting the forces of darkness.

Contrary to what is usually said, Peter's argument is well founded. In 1882 he questions whether *simply looking* at the sheep vaccinated at Pouilly-le-Fort can show that there is a *general* method, applicable to *all* infectious diseases. He calls this a "hasty generalization." Nor does he want to put an end to discussion by heroicizing Pasteur. At the Académie he cries out: "As for the term 'wonderful' that you use to describe the experiment of Pouilly-le-Fort, it is no longer an apologia, but an auto-apotheosis and I do not wish to have any part in that" (1883: pp. 558, 560).

How can one deny that he is right here? Peter does not want to turn a scientific experiment into a miraculous, divine event and to be extended without proof to every disease. Was not scientific method on his side? And yet he was wrong, if not for the reason we might think. He imagined that he was fighting against a scientist, whereas he was fighting against someone who was already the spokesman, the figurehead, and the amplifier of an immense social movement that passionately wanted Pasteur to be right and therefore made sure that all his laboratory work proceeded with a "haste" and a "widespread application" that were truly "prodigious." Peter claimed that the king was naked, but others rushed up to clothe him. Peter fought bravely, but he miscalculated the balance of forces and was therefore to sink into ridicule.

Koch did not share the same weaknesses, and he attacked Pasteur far from Paris and on the terrain of the new scientific medicine. But his criticisms intersect with those of the "backward-looking" Peter. Pasteur, Koch claims, generalized much too quickly: "M. Pasteur had already given himself up to the most ambitious hopes. With utter confidence he announced a forthcoming triumph in the struggle against infectious diseases" (1883, p. 65).

Koch finds all this premature. The technical objections he raises give us some idea of how anxious everyone was to agree with Pasteur. We cannot, Koch claims, generalize from one animal to another, nor from animal to man, nor from one disease to another, nor from the vaccination of a few individuals to that of all individuals. Koch challenges Pasteur to show the complete stock on which is credited the general method that is about to eliminate all diseases and revolutionize medicine. No one can deny that in 1881 this stock was extremely limited. The immense trust in Pasteur derived partly from the work that he had done before 1871, which did not concern infectious diseases, and partly from the social movement that needed these discoveries but went well beyond them without waiting for them to be made. In order to create networks of sanitation and to increase the circulation of goods and people, general laws as well as safe roads were needed. Koch's precautions weakened and interrupted the networks that the hygienists wanted to extend and strengthen. The hygienists cared nothing for Koch's precautions. Their trust went entirely to the man who was enunciating a general law and a principle of indefinite extension of the networks that they were going to command.

The critiques by Peter and Koch force us to see the disproportion between the forces that supported the generalizations of the Pasteurians and the scanty proof that they could provide at the time. If I insist on this point, it is because the history of the sciences is seldom just to the defeated or even, for that matter, to the victors. It accords too much attention to the latter and not enough to the former. A juster approach would be to treat both victors and defeated *symmetrically*. When Richet writes in the *Revue* in 1886 about the suggestion made to the Académie to set up the Institut Pasteur—"We are assured that to propose a vaccinal establishment is already to announce its creation" (1886, p. 289)—we must understand what he says as anthropologists, for it is little more than a magical invocation. He is willing to hand Pasteur the keys of the Institut simply by suggesting its possibility. A century later Canguilhem, a French historian of science, takes up the same incantation when he writes about the *Mémoire* of Pasteur on the theory of germs: "This theory, which already carried within itself, through the work of Koch and Pasteur, the promise, which was to be fulfilled, of cure and survival for millions of men and animals to come, also brought with it the death of all the medical theories of the nineteenth century" (1977, p. 63)[20].

We must analyze these beliefs in the power of what is in germ *in the same terms* as when Koch, proposing a vaccine against tuberculosis at the International Congress of Medecine in 1896, is besieged by patients from all over Europe possessed of the hope of being cured. Richet's confidence is made up of the same "credibility" as the "credulity" of the patients.[21] The fact that Pasteur had indeed funded his Institut, whereas Koch had to withdraw his vaccine in confusion, should not mislead us. In both cases Koch and Pasteur were sustained by a wave of trust, which they used as much as the patients used them.

There Was a Traitor among Us

So the hygienists translated this great conflict between wealth and health, without which their views would have interested nobody. But because they acted in every direction, their views remained in dispute and were little obeyed. Their various projects of sanitation were constantly interrupted by what seemed to them to be the ill will of other agents. They attributed all these diversions and decelerations to three kinds of ill will—first, to inertia on the part of the public authorities, who did not do what they ought to do; second, to what we would now call the "sociological resistance" of the masses, ignorant of their own interests; and last, to those diseases that appear and disappear, whose unworthy behavior is called "morbid spontaneity." In fact, these three kinds of resistance were connected. The hygienists' inability to prevent the outbreak of disease justified in advance the inertia of others. In order to mobilize the public authorities and, indirectly, the inert masses, they needed to be able to drive a sanitized path through the cities that no agent could interrupt or divert. But this was never the case.

A salesman sends a perfectly clear beer to a customer—it arrives corrupted. A doctor assists a woman to give birth to a fine eight-pound baby—it dies shortly afterward. A mother gives perfectly pure milk to an infant—it dies of typhoid fever. An administrator regulates the journey of Moroccan pilgrims to Mecca—cholera returns with the sanctified pilgrims and breaks out first at Tangiers, then at Marseille. A homemaker takes on a Breton girl to help the cook—after a few months the cook dies of galloping consumption. We always think we are doing the right thing, but our actions never turn out as we expected and are slightly diverted from their aim. The tribunal punishes a criminal with one year's imprisonment, but he pays for his brief spell in the cell with his life. When a man follows a woman to her hotel, he thinks he is settling the transaction with a coin and ends his days in an asylum. This displacement of the best-intentioned actions is truly discouraging: "For what I do is not the good I want to do; no, the evil I do not want to do—this I keep on doing" (Rom. 7:19).

But the situation is even more discouraging in that this distortion does not always occur. A lot of beer arrives intact at the retailers; many of those who frequent whores do not become syphilitics; many midwives do not kill their clients' babies. It is precisely this variation that is disturbing. It is the impossibility of predicting the intervention, the parasitism, of other forces that makes the remedies and statistics of the hygienists both so meticulous and so discouraging. Sometimes cholera passes, sometimes not; sometimes typhus survives, sometimes not. Indeed, the doctrine of "morbid spontaneity" was the only really credible one. Between the act and the intention is a *tertium quid* that diverts and corrupts them, but it is not always present, and we cannot capture it without taking everything into account at once: the heavens, weather, morals, climate, appetites, moods, degrees of wealth, and fortune.

This corruption of the best intentions, a corruption that was all the more disturbing in that it did not always occur, had one serious inconvenience. It encouraged skepticism. Steps could be taken, of course, but against what? Against everything at once, but with no certainty of success. It was difficult to arouse enthusiasm and sustain confidence in programs of reform and sanitation that all rested on this inconstant constant: "Confronted by this periodically recurring fatality, we remained powerless, unarmed, and, as the poet has it, 'weary of all, even of hope' " (Bouley: 1881, p. 549).

The skepticism led straight to fatalism. Indeed, this corruption of intentions had altogether too much the character of the "corruption of this world" for it not to be seen as inevitable. Life and good health were miracles, and neither the hygienists nor the doctors had much to do with them. They might wish to sanitize and reform, but it was difficult to convince the public and the public authorities to invest enormous sums of money over decades if the simplest programs could be betrayed by a sort of fifth column that undermined them from within. We can see the paradox of the hygienist movement: on the one hand, it was a social movement of gigantic proportions that declared itself ready to take charge of everything, and on the other, it was a succession of measures that were being quietly undermined by unknown and erratic agents. As a result, the period showed keen interest in identifying the corrupting forces, the double agents, the miasmas and contagions, and accorded immediate trust to those who might, in identifying them, be able to take measures against them. It was at this precise point that the microbe and the revealer of microbes appeared.

Between the beer and the brewer there was something that sometimes acted and sometimes did not. A *tertium quid:* "a yeast," said the revealer of microbes. When you send out the beer, you send out the barrel, the liquid, the delivery documents, *and* the yeast (Tyndall: 1877, pp. 789–800). When you bring a woman to birth, you think you are in the presence of three agents—the midwife, the baby, and the mother—but a fourth takes advantage of the situation to pass from your hands to the woman's wounds. Your interest is the life of the woman, but the interest of that fourth agent is different. It uses your interest to carry out its own. It proliferates; the woman dies; you lose a client (Duclaux: 1879, pp. 629–635). You organize a demonstration of Eskimos in the museum. They go out to meet the public, but they *also* meet cholera and die. This is very annoying,

Impact Microbes must be considered

because all you wanted to do was to show them and not to kill them (Anon.: 1881, pp. 372–377). Traveling with cow's milk is another animal that is not domesticated, the tuberculose bacillus, and it slips in with your wish to feed your child. Its aims are so different from yours that your child dies.

In order to understand what constituted Pasteurism up to the end of the century, we must understand what the Pasteurians, few in number, offered the hygienists. Working in few laboratories, they pronounced words that were immediately regarded as truthful and were integrated into evidence that at last allowed the hygienist movement to get on with its work. The hygienists were not "credulous." They *expected* something important from Pasteurism, something even more important in that they had been so disappointed before and were now sustained by a wider social movement. The small group of Pasteurian researchers created neither medicine, nor the huge body of theories on the causes of epidemics, nor the statistics, nor the determination of the social body to sanitize and remodel itself, nor even the rapid understanding by others of what they said. Yet they added something of their own, something that seemed essential to those who adopted it in order to pursue their own projects of sanitation.[22]

If we could go back to this impossible state of hygiene *before* Pasteur came to be credited with the whole movement, his contribution might be defined as that of a fulcrum. The Pasteurians provided neither the lever nor the weight nor even the worker who did the work, but they provided the hygienists with a fulcrum. To use another metaphor, they were like the first observation balloons. They made the enemy visible. Without replacing the armies, the battles, or even the commanding officers, they indicated or directed the blows. They were both nothing and everything. Duclaux, speaking of the surgeons who were the first to adopt Pasteurism, puts it well: "Surgeons have long proved that they have the noble ambition of doing good, whatever trouble it takes them, and they only had to be shown where the enemy was for them to learn to rush at those infinitely small enemies that had so often robbed them of their success and glory and that it was to be the honor of our century to have learned to know and to confront" (1879, p. 635). The Pasteurians were to displace (or translate) the intentions of the hygienists by adopting their projects, while adding to them an element that would *strengthen* both the hygienists and the Pasteurians.

There Are More of Us Than We Thought

We do not know who are the agents who make up our world. We must begin with this uncertainty if we are to understand how, little by little, the agents defined one another, summoning other agents and attributing to them intentions and strategies. This rule of method is especially important when we are studying a period when the number of agents was suddenly multiplied by millions. What struck all the authors of the *Revue* may be summed up in the sentence: "There are more of us than we thought." When we speak of men, societies, culture, and objects, there are everywhere crowds of other agents that act, pursue aims unknown to us, and use us to prosper. We may inspect pure water, milk, hands, curtains, sputum, the air we breathe, and see nothing suspect, but millions of other individuals are moving around that we cannot see.

"Ignoring the danger of the microbe awaiting us, we have hitherto arranged our way of life without taking any account of this unknown enemy" (Leduc: 1892, p. 234). Everything is in that sentence. There are not only "social" relations, relations between man and man. Society is not made up just of men, for everywhere microbes intervene and act. We are in the presence not just of an Eskimo and an anthropologist, a father and his child, a midwife and her client, a prostitute and her client, a pilgrim and his God, not forgetting Mohammed his prophet. In all these relations, these one-on-one confrontations, these duels, these contracts, other agents are present, acting, exchanging their contracts, imposing their aims, and redefining the social bond in a different way. Cholera is no respecter of Mecca, but it enters the intestine of the hadji; the gas bacillus has nothing against the woman in childbirth, but it requires that she die. In the midst of so-called "social" relations, they both form alliances that complicate those relations in a terrible way.

I am not using the word "agent" in any metaphorical or ironical sense but in the semiotic sense. Indeed, the social link is made up, according to the Pasteurians, of those who bring men together and those who bring the microbes together. We cannot form society with the social alone. We have to add the action of microbes. We cannot understand anything about Pasteurism if we do not realize that it has reorganized society in a different way. It is not that there is a science done in the laboratory, on the one hand, and a society made up of groups, classes, interests, and laws, on the other. The issue is at once

Make Room for Microbes

much more simple and much more difficult. To make up society with only social connections, omitting the invisibles, is to end up with general corruption, a perverse deviation of good human intentions. In order to act effectively between men—that is, to go to Mecca, to survive in the Congo, to bring fine, healthy children to birth, to get manly regiments—we have to "make room" for microbes. As Leduc puts it: "Science began the enslavement of the forces of nature and placed at the service of modern societies workers in iron and fire more powerful than all the slaves of the ancient world. But no science imposes as hygiene does interdependence on human societies; today we know that it is more or less impossible to benefit from the good things that it offers if we do not extend them to our neighbors; in other words, individual hygiene is closely dependent on public hygiene; a single unhealthy house in a town is a perpetual threat to all its inhabitants; if we are to give those good things to one, hygiene requires that they be extended to all" (1892, p. 233).

With what does Leduc make up his world? With "science," with iron and fire machines, with enslaved forces, but also with contagious diseases. The juridical "social" link is weak, but that which links *all* men together by *a* disease is much stronger. So what can we say about the juridical link redefined by the hygienist that must act *everywhere* in order to make the whole social body interdependent?

The Pasteurians redefined their numbers, with little regard to whether some belonged to nature and others "to culture," as the expression used to be. What interests them is whether they can be enslaved and what new forces can be created with these strange allies. Armaingaud, for instance, forms an odd alliance with the microbes: "In our struggle against phthisis . . . we have at our disposal an element of success that is largely lacking in the struggle against scrofula and the local tuberculoses: it is the motive derived from personal interest, the contagion that makes us all interdependent upon one another, the rich as well as the poor, the strong as well as the weak" (1893, p. 37).

Armaingaud, a rather paternalistic reformist, uses the microbe to redefine that celebrated "self-interest" and to link everybody together through fear of disease. This unexpected strengthening is not in itself "reactionary," as suggested by some authors who are used to speaking only of power and who see hygiene as a "means of social control." The allies of the microbe are to be found on the left as well as on the right. At the time of the inauguration of a Pasteur Institute in New

York, Gibier writes: "Later we shall see that the study of contagious organisms, which must be at the scientific base of hygiene, can and must bring considerable assistance to those who, finding that all is not for the best in the best of all possible worlds, are trying to improve the wretched condition of the disinherited." He speaks of "assistance." He, like a vulgar Marxist, attaches bacteriology to the class struggle: "The wretchedness of the poor distills a bitter and virulent bile that reaches as far as the rich man's goblet and contaminates the veins of his children" (1893, p. 722). The poor may have no rights, but the contagious poor can blow up the whole outfit. What is refused to one cannot be refused to the other. The class struggle may be stemmed at one point, only to reappear at another through contagion. Rosenkranz shows in the case of similar reforms in the United States the impossibility of telling whether they served the right or the left because the microbe rendered unpredictable interests that would be too predictable without it[24].

No one, toward the end of the century, could do without contagion in connecting men, plants, and animals. In an article on the role of microbes in society, Capitan sums up his thinking: "I have just outlined the way in which pathogenic microbes evolve in society . . . Society can exist, live, and survive only thanks to the constant intervention of microbes, the great deliverers of death, but also dispensers of matter" (1896, p. 292). Again, it is as an anthropologist that we must follow these new translations of what matters in the world. Capitan distinguishes in a different way between what is beneficent and what is harmful, what is useful and what is useless, what acts and what does not. He does not base society on biology, like a vulgar contemporary; he redefines society itself, a society in which the new agents intervene now and at all points. "We need the assistance of the infinitely small," writes Loye (1885, p. 214). Microbes connect us through diseases, but they also connect us, through our intestinal flora, to the very things we eat: "We can hardly doubt the importance of the role played in the economy of the individual by those table companions that help it to break down organic substances" (Sternberg: 1889, p. 328).

"Interdependence," "assistance," "power," "help," "table companions"—I have not imposed these terms; they all emerge from the trials of strength. It is the actors that thus redefine their worlds and decide which must now be taken into account[25].

Must be willing to Sacrafice for others

From the Science of Society to the Study of Associations

The actors whom we are studying already have many lessons to teach us. In particular, they do not wait for the sociologist to define for them the society in which they live.[26] They reorganize society with new actors who are not all social. Sociologists of the sciences often claim to be providing a political or social explanation of the content of a science, such as physics, mathematics, or biology. But the sociology of the sciences is too often powerless, because it thinks it knows *what* society is made up *of.* Faithful to its tradition, it usually defines society as made up of groups, interests, intentions, and conflicts of interest. So we can see why this sociology is so feeble when it approaches the exact sciences. It thinks it can explain hard disciplines in social terms, whereas those disciplines are almost always original and more subtle *even in their definition of the social body* than sociology itself. Sociologists of science think they are very clever because they have explained hygiene in terms of the class struggle, the infrastructure, and power, whereas the agents spike our guns for us. They go off and look for new allies to advance the cause and to terrify the rich (or poor), brandishing diseases. Which explains the other? Which is the more inventive?[27]

The exact sciences elude social analysis not because they are distant or separated from society, but because they revolutionize the very conception of society and of what it comprises. Pasteurism is an admirable example. The few sociological explanations are feeble compared with the strictly sociological master stroke of the Pasteurians and their hygienist allies, who simply redefined the social link by including the action of the microbes in it.[28] We cannot reduce the action of the microbe to a sociological explanation, since the action of the microbe redefined not only society but also nature and the whole caboodle.

Microbes are everywhere third parties in all relations, say the Pasteurians. But how do we know this? Through the Pasteurians themselves, through the lectures, the demonstrations, the handbooks, the advice, the articles that they produced from this time. Who, then, was the third party in all these social relations at the time? The Pasteurian, of course, the revealer of microbes. For whom must we "make room"? For millions of omnipresent, terribly effective, often dangerous, and quite invisible microbes. But since they are invisible, we also have to make room for the revealer of microbes. In redefining the social link as being made up everywhere of microbes, Pasteurians and hygienists regained the power to be present everywhere. We cannot "explain" their actions and decisions by "mere" political motives or interests (which in any case would be very difficult to do.) They do so much more. In the great upheaval of late nineteenth-century Europe, they redefine what society is made up of, who acts and how, and they become the spokesmen for these new innumerable, invisible, and dangerous agents.

The lesson in sociology that Pasteurians and hygienists give to their time (and to sociologists of science) is that if we wish to obtain economic and social relations in the strict sense, we must first extirpate the microbe. But in order to extirpate the microbe, we must place the representatives of the hygienists or Pasteurians everywhere. If we wish to realize the dream of the sociologists, the economists, the psychologists—that is, to obtain relations that nothing will divert—we must divert the microbes so that they will no longer intervene in relations everywhere. They and their ways must be interrupted. *After* the Pasteurians have invaded surgery, only then will the surgeon be alone with his patient. After we have found a method of pasteurizing beer, then the brewer will be able to have nothing but economic relations with his customers. After we have sterilized milk by spreading throughout all farms methods of pasteurization, then we will be able to feed our infant in a pure loving relationship. Serres describes this elimination of a parasite by another more powerful one.[29] Only after the insulation of the second parasite can we declare ourselves safe from the first. At the cost of setting up new professions, institutions, laboratories, and skills at all points, we will obtain properly separated channels of microbes, on the one hand, and of pilgrims, beer, milk, wine, schoolchildren, and soldiers, on the other.

To explain bacteriology is not, then, to reduce Pasteur to the position of a social group. On the contrary, it is to follow the lesson that bacteriology and hygiene gave to all the sociologies of the period. "You thought you could do without the microbe. Yet the microbe is an essential actor. But who knows it? We, only we in our laboratories. So you must take us into account and go through our laboratories if you are to solve the problems of society." In order to understand this point of view, we must remember that the period was full of people who turned themselves into the spokesmen for dangerous, obscure

Declare one safe from the other

forces that must now be taken into account. The hygienists were not alone in inventing new forces. There were those who manipulated the fairy electricity, those who set up leagues for colonization, for the development of gymnastic clubs, or for promotion of the telephone, radio, or X-rays. The radical party, for instance, gained ground everywhere by forcing the traditional agents of the social game to take account of the dangerous laboring classes, whose actions and intentions were so little known. But it is with Freud that the resemblance is greatest. Like Freud, Pasteur found treasure, not in the parapraxes and trifles of everyday life, but in decay and refuse. Both announced that they were speaking in the name of invisible, rejected, terribly dangerous forces that must be listened to if civilization was not to collapse. Like the psychoanalysts, the Pasteurians set themselves up as exclusive interpreters of populations to which no one else had access.

It does not matter that Pasteur developed an exact science, that the radical party occupied a growing place in parliament, and that Freud developed a science that is still controversial. It does not matter that some define human actors and others define nonhuman agents. Such distinctions are less important than the attribution of meaning and the construction of the spokesmen who express, for others' benefit, what is being said by the unconscious, the rabies virus, or the print worker. Such distinctions among types of actors matter less than the fact that they are all renegotiating what the world is made up of, who is acting in it, who matters, and who wants what. They are all creating—this is the important point—*new sources* of power and new sources of legitimacy, which are irreducible to those that hitherto coded the so-called political space. They cannot be reduced to a "social or political explanation," since they are renewing the political game from top to bottom with new forces. If socio-logy wishes to be the science of "social facts," then it cannot understand this period. It thereby limits itself to the purely social, whereas all the actors are dirtying it with something else. More seriously, sociology remains deaf to the lessons of the actors themselves. If we wish to learn from this lesson and still call ourselves sociologists, we must redefine this science, not as the science of the social, but as the science of associations. We cannot say of these associations whether they are human or natural, made up of microbes or surplus value, but only that they are *strong* or *weak*.[30]

How to Become Indisputable

We begin to understand the general process of translation found in the *Revue Scientifique*.

"We want to sanitize," say the hygienists, expressing in their own way the forces of the period and the conflicts between wealth and health.

"All your good intentions are diverted, confused, parasitized," say their enemies.

"This parasite that diverts and confuses our wishes, we see it and reveal it, we make it speak and tame it," say the Pasteurians.

"If we adopt what the Pasteurians say, seizing the parasite with its hand in the bag, we can then go as far as we wish," say the hygienists. "Nothing will be able to divert our projects and weaken our programs of sanitization."

In spreading the notion of the Pasteurians as revealers of microbes, the hygienists, who claimed to be the legislators of health, spread themselves. By generalizing both the Pasteurian and the hygienist everywhere, the desire to get rich was no longer thwarted. The conflict between health and wealth was resolved to the benefit of the latter.

As McNeill suggests when discussing the millenium-long struggle between the microparasites and the macroparasites, a struggle that seems to him to be the motive force of history, the scale is turned in favor of the macroparasites.[31] The rich and the empires will at last be able to spread. Hitherto, especially in the tropics, they could never go very far. Their most faithful factotems soon died. Now, wherever the Pasteurians and hygienists gained ground, the microparasites lost ground. We can see why nobody, even today, can seriously question the contribution of bacteriology. Indeed, all opinions speak with the same voice, and everybody works together to attack the microparasites: exploiters, exploited, benefactors of mankind, merchants, the clergy, and above all the doctors, the hygienists, the army medical corps, and at the end of the parasitical chain, the Pasteurians. The only losers in all this are the microbes. Since no human being can wish to defend them, the general transformation of towns in the nineteenth century through the elimination of microbes is indisputable.

The assemblage of forces that I am trying to reconstruct might be confused with the final impression given by this assemblage if it were not for a certain distinction. Microbes might have been discovered, but with no responsibility for this discovery being attributed to Pasteur. After all, in politics, ingratitude is more common than gratitude. Thus, two mechanisms must be distinguished. The first sets up the forces one on top of the other and enables us to explain how a whole period is interested (finds itself interested) in what is happening in Pasteur's laboratory; the second mechanism attributes responsibility for the command to one member in the crowd. When Tolstoy explains the Russian campaign, he describes the first mechanism, but he is well aware that the second is constructed differently, since the maneuvers are attributed to "Napoleon's genius" and "Kutuzov's genius." The same goes for bacteriology. What I call the primary mechanism shows how bacteriology got into the end of the parasitical chain and found itself able to express a whole period. But the secondary mechanism attributes the whole of the sanitational revolution of the period to Pasteur's genius. The primary mechanism describes the alliances and make-up of the forces, whereas the second explains why the forces are mixed together under a name that represents them. The first defines the "trials of strength"; the second enables us to explain what "potency" is made up of.

This is not a minor point, for it helps us to explain two very different things: first, how the hygienists or Pasteurians put themselves in a position to translate the forces that needed them and, second, how they initiated an investigation to define who was responsible for the movement as a whole. I have said that the shift took place only through translation. But this translation is always a misunderstanding in which both elements lay different bets. Once the shift has been made, it is crucial to decide who was ultimately the cause of this transaction. For instance, it is almost certain that the English bacteriologists arranged their laboratories in the same way as the French biologists. So the primary mechanism was the same. But it was only in France that responsibility was attributed entirely to bacteriology, which was reputed to be the work of a single man, Pasteur.

This distinction between the two mechanisms is an essential one, because the strategies that it implemented were quite different and could vary in the same article. For example, Richet, speaking of the antidiphtheria serum, ends: "It may be astonishing that I have not seen fit to mention the great name of Pasteur. But what is the point? Do we not know that every discovery in the domain of bacteriology emanates directly from M. Pasteur, just as every discovery in chemistry emanates from Lavoisier?" (1895, p. 69). Plotinus himself would not have endowed his God with enough power to make the antidiphtheria serum (to which Pasteur the man hardly contributed) emanate "directly" from him. But the same Richet, in the same article, uses a quite different model to establish the priorities of the discoveries about diphtheria: "In the presence of this magnificent result, this victory over death by science, it is of relatively secondary interest to know to whom it is due; for we always exaggerate what is attributable to a particular scientist in any discovery. It is, much more than his pride supposes, due to an anonymous, perpetual collaboration and to the exchange of ideas in the air, each of which makes its useful, obscure contribution" (p. 68).

What? So there are "anonymous" researchers? Other researchers than Pasteur? "Collaborations," "ideas in the air"? The Plotinus-like emanation has become the humdrum sociology of the sciences, a crowd of anonymous, hard-working foot sloggers. One may have guessed the cause of this shift in metaphysics. The first obscure, anonymous collaborator was none other than Richet himself: "On December 6, 1890, we carried out the first serotherapic injection on a man" (1895, p. 68). This double game of explanation—one creating potency, the other setting out the trials of strength—might seem no more than an amusing oddity. But it helps us to explain how so patent a manipulation of all the trials of strength of a society may end up giving the impression that a society has been revolutionized by the purely scientific ideas of a few men, and it even helps us to explain how, by reduction to the secondary mechanism, we end up with the impression that there exists a science on the one hand and a society on the other.[32]

See Note Science + Society Separate? is an artifact

Hygiene and the Obligatory Points of Passage

Let us take a look at the side of the hygienists and see why they seized so readily on any argument about microbes to emerge from the microbiological laboratories. I said that they were at war and were fighting on all fronts. I compared them to a small army given the task of defending an immense frontier and therefore obliged to disperse itself along a thin cordon sanitaire. They were everywhere, but were everywhere weak, and we know how many epidemics, how many outbreaks of typhus, cholera, yellow fever, got through those ill-defended frontiers. What does the definition of the microbe and the description of its habits mean to them? Precisely what in the army are called obligatory points of passage. Depending on its equipment, the enemy cannot get through everywhere, but only in a few places. They have only to concentrate their forces at those points for their weakness to turn into strength. The enemy may then be crushed.

Take an infantile disease like the ophthalmia of the newborn, a cause, say the statistics, of 30 percent of those born blind. Fuchs, the author of this article, says that he believed like others in morbid spontaneity. Anything could cause this ophthalmia: overbright light, cold, jaundice. Then he adds, without actually citing Pasteur: "As soon as we learned that the cause of several infectious diseases lies in microscopic mushrooms, we were all ready to believe that here, too, the blame lay with the microorganisms" (1884, p. 494).

The mere definition of an agent is enough "to lead us to believe"— a crucial term—in a new program of research. Thanks to the prowess of this agent, Fuchs sets about linking two hitherto unconnected statistical aggregates, the presence of disease and gonorrhoea in the mother. He then finds the same gonococcus in the mother's wounds and in the puss discharging from the infants's eyes. When could the microorganism pass from the mother's vagina to the well-closed eye of the newborn infant? There was only one answer: through the lashes to which it adhered. This was the obligatory point of passage: the eyelashes. But where does Fuchs strike? In the eyes themselves. With what? With a powerful disinfectant, silver nitrate. Fuchs was powerless to prevent all the causes of a disease. He found himself in a strong position crushing the gonococcus with silver nitrate at the single place where it was obliged to pass. The results of this new trial of strength were spectacular, "indisputable." In a German hospital, says the author, the figures dropped from 12.3 percent of diseased children to 0 percent. Who indeed could still argue about that? By deploying the same forces, Fuchs gets results that bear no comparison with earlier ones. Understandably, this reinforcement is enough to show why so many people were "led to believe" in the presence of microbes.

Furthermore, the microbe made it possible for a reordering of epidemiological problems, where it seemed that the number of causes would always defy analysis. Take, for instance, an investigation into a cholera case at Yport, a little harbor in Normandy. The investigator, a certain Gibert, is confronted by a puzzle worthy of Sherlock Holmes: a Newfoundlander lands with his fish at Sète in the south of France; a sailor dies at Toulon; in the train a bag belonging to the dead man travels unaccompanied; at Yport a woman washes the linen of her sick brother; she lives in a steeply rising street; there is a public fountain. From the only point of view known to statistics, this miscellany of disparate facts can only produce the following: in 1884 there was an outbreak of cholera at Toulon and another at Yport in Seine Inférieure. "The doctrine of morbid spontaneity has been mentioned once again," the author admits.

To remedy the uncertainty, the investigator comes on the scene with his Ariadne's thread. He imposes his preconceived certainties on the investigation. The microbe is not an idea floating in the head of scientists; it is a means of locomotion for moving through the networks that they wish to set up and command. The microbe is a means of action, designed for a certain use and a certain type of connection and movement. There is a specific microorganism; it does not jump from one place to another; we must follow the thread. With these certainties, a new route is both described and dug. Gibert recounts how the sick sailor, a friend of the sailor who died at Toulon, has his linen washed by his sister: "The day after his arrival, he had all his clothes soaked, in two lots, in a tub, then had them hung up to dry . . . The water from the tub was thrown out into the very steep street and traveled over 50 meters" (1884, p. 724). Seven died along that steep street! "Each new case could be connected with the earlier cases and there was not a single one that was not explained by contagion" (p. 725).

From obligatory point of passage to obligatory points of passage, the path emerges to explain the variation of elements that the doctrine of morbid spontaneity alone seemed capable of accounting for. Contagionism as a general doctrine was powerless, but the Ariadne's thread, making it possible to connect a ship, a train, a particular topography, a system of water supply, brought together both the traditional investigation and the new agent. Before, *everything* had to be taken into account, but in a disconnected fashion; now the hygienist could also take everything into account, but *in the order laid down by the* microbe's performances. It is easy to imagine the extraordinary enthusiasm of all the hygienists called upon to discover the traces of an enemy that seemed so erratic as to summon up the whole explanation of morbid spontaneity. Without abandoning anything of the past, they were becoming stronger. "If we could know the microbe at the source of each disease, its favorite haunts, its habits, its way of progressing, we might, with good medical supervision, catch it in time, stop it in its tracks, and prevent its continuing in its homicidal mission" (Trélat: 1895, p. 169).

I have chosen on purpose three authors who were not Pasteurians, did not mention Pasteur, or worked on diseases that the Pasteurians had not yet dealt with. Indeed, the formidable transformation of hygiene was effected at first only with the following research program: there are obligatory points of passage; the microbe is the Ariadne's thread that links all the points together. Of course, we may admit that Pasteur was responsible for the certainty that specific microbes existed, but he was not responsible for their medical use.

The clearest case was obviously that of surgery. In the *Revue* its transformation is regarded as won from the outset. Indeed, from the point of view of the secondary mechanism surgery is regarded as the work of Lister and Guérin. Pasteur is seen, at least at the beginning, only as the occasion of a development for which the surgeons themselves were responsible. We understand why. Antisepsia and asepsia may develop without the knowledge of any particular microbe, without the culture, the attenuation in short, without anything to be found in the medical program of the Pasteurians. In order to launch Lister, all that was needed was for surgeons no longer to question the existence of microbes and their ability to pass everywhere, but for them to know more or less that microbes died in heat or in the air—or absence of air—under the effects of a disinfectant. All Pasteur had to do was to make this indisputable, and the surgeons themselves would "apply" it, that is, do the rest.

The enthusiasm of the surgeons shows clearly enough that we cannot distinguish "belief" from "knowledge." The degrees that lead from the most skeptical indifference to the most passionate fanaticism are continuous and measure the angle of relations between the agents. We believe that which we expect something from in return; in this sense belief is based, like knowledge, on the extension of safe networks that allow things to go and come back. For instance, asepsia allowed the surgeons to reach new places that they were unable to reach hitherto except on corpses. Thus their beliefs, their knowledge, and their skills grew at the same pace and in the same proportion:

> The act of operating no longer kills: we are more or less masters of the cuts we make, we direct them almost at will toward immediate healing . . . The serious interventions of former times, the amputations of limbs, the hollowing out of bones,

articular resections, removal of breasts, first entered everyday practice. Then the horizon widened: abdominal surgery was created out of nothing. We cut, we resected, we sewed up the stomach, the intestines, the liver and its biliary vesicle, the spleen, the kidney, the pancreas itself . . . Antisepsia made this miracle possible: complications in wounds were now the exception, and thanks to M. Pasteur's discoveries, M. Lister has deserved the celebrated gold statue promised by Nélaton to whoever delivered us from purulent infection. (Reclus: 1890, p. 104)

It is not a question of ideas, theories, opinions. It is a question of ways and means. Surgeons could go into the stomach, they could wield the lancet in the ovaries, and still hope that the living individual on whom they were operating would not die at once. The certainty that surgeons placed in the antiseptic method corresponded exactly to the territories that it was opening up to them. The translation appears quite clearly. At the cost of a rapid and inexpensive detour via the gestures of disinfection, they reached more quickly and further to what they had been wishing to reach since antiquity.

I have already cited the unfortunate Kirmisson, lamenting the powerlessness of surgeons to control at once all the factors of purulent infection: "So we had accumulated all the precautions of general hygiene, but we had not managed to uproot from the wards purulent infection and all the calamities of surgery." Such was the first program, the first hygiene. He adds: "We were obviously on the wrong track; we were looking for the cause of the accidents in the environment, in the hygienic conditions in which we found the patient, whereas we had to fight them and above all prevent them by the use of antiseptic substances in the wound itself" (1888, p. 296). The reversal was made. The wound was enough: it was there that surgeons had to take precautions. The environment was of course important, but they will never be strong enough to control it entirely. The weak became strong simply by changing the point of application of their efforts. Protect the wounds and not the environment: that was enough to redirect the forces of surgery as a whole, which became almost at once stronger than the microbes that were perverting their good intentions.

This transformation may be expressed more precisely. The surgeons passed from a total attack to a specific attack, or in other words, from a full totality to a *hollowed out* totality. Before there was endless discussion in the *Revue* about "disencumbering" the hospitals. This solution was typical of the "old" hygiene with its precautionary accumulatory methods. There are too many men, too many diseases, especially in the cities. They must be cleaned out. But Kirmisson writes again, if all we have to do is to protect the wounds, "the question at the present time is oddly simplified . . . We are no longer demanding, as in the past, that the old hospitals be pulled down and new ones built at a cost of millions. These great hospitals, imperfect as they are, from the point of view of general hygiene, are adequate to our needs providing we practice strict asepsia" (1888, p. 297).

Not only surgery was simplified and strengthened, but hygiene as a whole, which could vindicate its advice by concentrating its forces on the obligatory points of passage. It was not always convenient to follow Bouchardat's prudent advice to physicians assisting at births to wait several days before doing so again. It was simpler to wash their hands in a carbolic lotion after each childbirth (Bouchardat: 1873, pp. 552–564). It was expensive and ineffective to build maternity homes. Yet it was quite possible to place the women in close proximity, providing they were surrounded by an antiseptic cordon sanitaire (1875). Quarantine is an inconvenient method. Why lock people away when you let their infected linen escape? Jousset de Bellesme asks indignantly (1876, p. 403). They must simplify the precautions to be taken. When ten years later it was discovered that cholera had only a five-day incubation period, the quarantine could safely be reduced to six days. There was controversy about the danger of cemeteries. But since no passages were found to link the microbes of the dead with the living, they could declare cemeteries healthy (Robinet: 1881, pp. 779–782). The same went for drains. Their smell was pestilential, but if microbes did not pass with the smell, they presented no danger.

Thus, all the great problems of hygiene—overcrowding, quarantine, smells, refuse, dirt—were gradually retranslated or dissipated. Either the microbe gets through and *all precautions are useless,* or hygienists can stop it getting through and *all other precautions are superfluous.* The hygiene that took over the doctrine of microbes became stronger and simpler, more structured. It could be both more flexible—quarantine could be relaxed—and more inflexible—total disinfection to 120 degrees. In a sense hygiene lost ground, since it was no longer directed at the totality, but in another sense it gained ground at last by striking more surely at an enemy that had become visible. This is why the contribution of the hygienists is difficult to isolate from that of their allies. They changed what they wanted to do, while at the same time achieving it at last by following Pasteurism. They were like people who had begun to set up a road network consisting of thousands of country lanes in order to travel everywhere and ended up building only a few main roads. The aim is still the same, to get everywhere, but the program of public works is quite different.[33]

Attribute responsibility to certain actors

The Hygienists Made Their Own Time

This shift in hygienic precepts, which became rarer and more firmly based, was also to transform the relationship between hygiene and its own past. Hygiene's style was cumulative and precautionary, since it set out to embrace everything. Lycurgus and Hippocrates were invoked by writers obsessed by the fear that, in ignoring one detail, they might be ignoring one of the causes of those diseases that have so many causes. As soon as they redeployed their forces, eliminated a lot of knowledge, and structured the advice available around the obligatory points of passage, they could ignore a large part of the opinion of the ancients and drop whole areas of what by this time had become "traditional" hygiene. After 1880 the style of the hygienists could be recognized at a glance. Once they had given their advice on everything; now they decided on a few things. Once they had accumulated everything; now they ordered. Time no longer moved in the same direction. Instead of advancing without moving and keeping everything, they retrenched, jettisoned, and as a result felt they were making progress at last.

Often in history when we see such differences of style or thought, we speak of revolution (borrowing the language of the politicians), or even of epistemological break (this time borrowing the language of the butcher's shop). But to explain even a radical difference by a break "in time" is to explain nothing at all. It is to suppose that time passes and dates exist. We always say, for instance, that time is irreversible. This is easily said. The year 1875, we claim, is after 1871. But it is not necessarily so. The hygienists always complained that things were not moving forward, or even that they were getting worse. For them, certain things had not changed since Galen. Is time irreversible? Would that it were! On the contrary, it is reversible—so reversible that it is possible not to have made any progress since the time of the Romans. Now if things stagnate, we can hardly make a distinction between 1871 and 1875, except on the calendar, which does not amount to very much.

In other words, it was only recently that hygienists had come to see the difference in years. Before, they could not decide. A particular piece of advice might have been archaic, but it might be useful tomorrow. A particular remedy was new, but it might only be a method that would be superseded tomorrow. Nothing was really without a future, but nothing really had a future. In such a state, nothing could divide up the time of hygiene into recognizable periods. Or rather some hygienists tried to do so, but in great epochs: "Theocratic with Moses, patriotic with Lycurgus, naturalistic with Hippocrates, metaphysical with the alchemists, it was only in the late eighteenth century, at the instigation of the Royal Society of Medicine, that [hygiene] became experimental, that is to say, truly scientific, resting on the biological and sociological sciences" (Corlieu: 1881, p. 533).

These spaces of time are not enough to distinguish between 1871 and 1875! In any case, the agents were not in agreement as to the date at which things began to change. For Martin in 1880, the "new era" began in 1876 at the Brussels Congress (1880, p. 1071). For Bouley in 1881, the foundation of hygiene dates, we are hardly surprised to learn, from Pouilly-le-Fort. How are we to distinguish between the years and how are we to produce a better periodization? This is the same problem that each actor has to confront.

If the years are to be distinguished from one another and if time is to go in one direction, we must create irreversible situations. There must be certain things that we can no longer go back on. Time—that is, the distinction between moments—is the distant *consequence* of actions to make a particular position durable. It is not, nor can it be, a cause. But for the agents to make their positions durable and irreversible, they need recognized properties, that is, achievements. This is why they threw themselves with such enthusiasm on "Pasteur" and always said that "his principles were so strongly established that they could never be overthrown." But they were the ones who did not wish to be overthrown, and that is why they made the principles so indestructible: the first turn of the ratchet. Noson, an unhealthy, stinking town, was "superseded" and "anachronistic." In the age of progress this was another turn of the ratchet. The statistical results of those efforts were uncertain, but with new methods the results at last had the unquestioned certainty of the sciences of time: before, so many deaths; after, none—here millions of microbes, there none. A third turn of the ratchet. The achievements were piling up. It was already becoming more difficult to reverse them. If the hygienists managed to recruit enough allies, then they would be able to make time irreversible. Then they would be able to begin to date years. Time, that celebrated time, would at last be able "to climb the ladder," as Péguy says in *Clio*.[34] Hence their enthusiasm. What they refer to as the recent event is a change in the *regime* of time: before, this regime did not move forward; now it does. Before, the hygienists could not, without being immediately contradicted, tell others what the time was and yell, "You are archaic and superseded." Now they could do so and no one would contradict them. Now they could do so *because* they were no longer controversial. This closure of the argument was due, in turn, to the allies that they gave themselves in order to make their positions impregnable.

To speak of "revolution" is difficult enough in politics, but it is impossible in such a subject. The temporal framework itself is useless. What makes the history of the sciences—so respectable elsewhere— usually disappointing is that it sets out from time in order to explain the agents and their movements, whereas the temporal framework merely registers after the event the victory of certain agents. If we really wanted to explain history, we would have to accept the lesson that the actors themselves give us. Just as they made their societies, they also made their own history. The actors periodize with all their might. They give themselves periods, abolish them, and alter them, redistributing responsibilities, naming the "reactionaries," the "moderns," the "avant-garde," the "forerunners," *just like* a historian— no better, no worse. We ought to ask history to display the same humility that we have asked sociology to do. Just as we asked sociology to abandon its "social groups" and its "interests" and to allow the actors to define themselves, we ought to ask history to abandon its "periods," its "high points," its "development," and its "great breaks." Nothing would be lost by this, for the actors are just as good historians as sociologists. Something would surely be gained by this: instead of explaining the movements of the actors by time and dates, we would explain at last the construction of time itself on the basis of the agents' own translations.[35]

The hygienists did not become modern after centuries of stagnation. They made themselves modern by *bypassing* all the others. It was to achieve such a supersession that every argument on the microbes was immediately seized upon, amplified, generalized, popularized, believed by those who had taken responsibility for directing the sanitization and regeneration of Europe. Their most advanced aims had become almost indisputable. At the cost of accepting microbiology, the support of its laboratories, and even the continual praise of the "great Pasteur," they advanced their cause more quickly and strengthened their positions everywhere by weakening their adversaries, whether microparasites or public authorities. The time that they *made* was now working for them.

We Must Know How to Bring a Science to an End

The only way by which the hygienists could make their achievements irreversible was to link the fate of what they were doing to something else that was less disputable. The hygienists paved the way for the Pasteurians by trusting them and generalizing what they said. They went further: they considered early on that microbiology was a complete, definitive science and that *all that remained* was to apply it. The first marker of this closing operation is to be found in the *Revue* in 1883: "From the day when the theory of parasites threw light on the hitherto still mysterious etiology of infectious diseases, we had to find out whether it would help in discovering the true nature of the malarial poison" (Richard: 1883, p. 113).

From now on, the certainty of the theory of parasites was taken as a premise either of research programs that *had only to be* implemented or of practical measures that had only to be applied or generalized.

We cannot explain this closing operation by saying that microbiology was at the time an exact science. Indeed, the exactness of a science does not come from within. It, too, comes from the strength of the agents with whose fate it has managed to become linked. Astonishing as the results already accumulated at the period by Pasteur may appear, they could not in themselves explain the trust accorded them by the other actors, for the excellent reason that the actors in question were the only ones to see what could be done with them. Nor can the scientism of the period alone explain this immediate trust, for controversies were just as passionately fought out in that century as in any other. The reasons there were no controversies should be, in order to respect the principle of symmetry, the same as those that opened them. If hygienists had wanted to open up a dispute, they could have done so. The absence or presence of a controversy is a measure only of the angles of movement of the actors. This is proved by the fact that Koch or the doctors of the *Concours Médical* opened up controversies on the very same objects, which seemed to the hygienists to be irreversibly closed.

Even if we admit that the content of microbiology was responsible for the trust placed in it, we cannot explain this closing operation in every case by the "real" efficacy of microbiology or of hygiene, since this operation *precedes* and *makes possible* the generalization of these two sciences. The best proof of this shift is provided by an editorial of 1889. The latest figures for deaths from infectious diseases, says Richet, "have been in constant progress for the past twenty-five years" (1889, p. 636). These figures ought also, if science and societies were Popperian, to put into question all the efforts of the hygienists and, still more, those of their Pasteurian allies. But this is not at all the case. Richet goes on: "We should not conclude from these figures that the efforts of the hygienists have proved useless or the achievements of science fruitless." Anthropologists have shown that, in a witchcraft trial, there are always agents that cannot be made responsible and others toward whom accusation invariably moves.[36] This is in fact what allows witchcraft to reveal so well the fabric of society. It is the same here. Doubt moves not in the direction of science but toward the inertia of the public authorities. As Richet continues, "Despite the progress of science, despite the advance of physicians and engineers, the hygienic economy of a great and ancient city like Paris remains more or less out of control." When it is a matter of forming alliances that are durable enough to overthrow the whole of urban Europe, no counterexample will prevail against these certainties, no accusation will be pointed in the direction of science or the Pasteurians. Such statements are measures not of the partiality or the credulity but of the capital of trust that had been invested in research concerning the microbes.

Here we can see that "trust" is never a primary term. It depends on the scope of the operations into which the hygienists threw themselves. Indeed, it is in the very nature of the transformations that they advocated to have no result until everything is finished. A single microbe may endanger everything. The hygienists are powerless to jugulate the infectious diseases if they do not invest continuously for several generations. To create irreversibility and to rid themselves forever of the microbe, they must not abandon the building of drains on the way or even suspend vaccination for a time. In no circumstances

Continuous

must they interrupt the disinfection of midwives' hands or the sterilization of milk. The network of gestures and skills that the hygienists wanted to set up had to be as continuous as an oil pipeline. Because they had to create that long time, the *Revue*'s authors cannot indulge in the slightest dispute concerning Pasteur. This was the only way of creating the future conditions for the realization of an efficacious Pasteurism. This is why it is pointless to claim that Pasteur's discoveries were believed because they were convincing. They ended up being convincing because the hygienists believed them and forced everybody else *to put them into practice.*

Hygienist *Put everything into practice*

In order to make their position permanent, the hygienists had to set up the greatest possible "potential difference" between the "indisputable conquests" of science and the "ditherings of the public authorities." This was, for the hygienists, the only way of *setting up* the "scandal" and getting government to budge. This setup is obvious at a very early stage: "From the theoretical point of view, in short, hygiene has done its work; but it has not gone beyond; as far as practice is concerned, we are behind most of the civilized nations, even though we were ahead of them on the purely scientific terrain . . . Everything remains to be done as far as practical implementation is concerned; but the solutions are there and we only have to implement them with the utmost speed" (Rochard: 1887, p. 389).

We can see the extent to which the notions of "behind" and "implementation," so often used in the sociology of innovation, are the result of a strategy to get other authorities to move at last. No researcher in his right senses could claim that there was no more to learn in bacteriology by 1887 or even that it could be implemented medically!

This setup has nothing whatsoever to do with an "intellectual" confidence in Pasteur's results or a love of science. What the microbe and the transformation of microbiology into a *complete* science did was to make long-term plans of sanitization *indisputable*. They offered, literally, a real guarantee of municipal investments. How could the hygienists convince city councils to throw themselves, for instance, into a public drainage program if there were still any dispute "in high places" as to its harmlessness? However, as soon as the scientific argument was closed, they could guarantee the municipalities a good return on their investments. Rochard writes: "Civic hygiene has been the subject of innumerable studies and we know everything we need to know for it to be possible to proceed to the sanitization of unhealthy localities without fear of making mistakes or committing ourselves to unproductive expenditure" (1887, p. 389). The final guarantee, at the end of the chain, was to be found in the micrography and microbiology laboratories.

We now see why the hygienists placed so much trust in Pasteur, rejected all controversy about him, and generalized his results. This result was not necessary. But science had to be raised to the highest possible level if the present state and inertia of society were to appear in all their scandalous starkness. We also see why the hygienists made "Pasteur" totally responsible for the whole of this movement, a result that was also not absolutely necessary. They did not concern themselves with microbes out of politeness any more than out of a love of science. If the angle of their movements had interrupted that of the Pasteurians, we might still be waiting for a Pasteurian "revolution" (as has occurred in the case of the doctors).

But the hygienists could make the potential difference even greater, and make the inertia of the public authorities still more scandalous, by attributing hygiene itself to "Pasteur." Pasteur was not the one who arrogantly claimed the new hygiene as his own work. It was the hygienists who needed to turn "Pasteur" into the advocate of all their decisions. We may dispute the work of a hygienist; we could not dispute "Pasteur." If the secondary mechanism accorded so much place to "Pasteur," it was again because the hygienists wanted it that way. Since there were more of them and they were more influential, the proceedings instituted by Pasteur to apportion responsibility would have been lost if, by chance, they had not agreed.[37]

From the New Indisputable Agent to the New Authorized and Authoritarian Agent

A strengthened, structured hygiene had created a future for itself and, instead of hesitating, now spoke with a new authority in every sense of the word. This is apparent in Richet's editorial of 1883:

> Engineers know the art of engineering. Do they know what typhoid fever is? Do they know the meaning of the word contagion? Do they read the mortality statistics for Paris? Administrators know very well what the administrative regulations are, but do they know what is meant by infection, disinfection, contagion, and epidemic? It is vital that the public authorities remain

deaf no longer to the appeals of the hygienists, which become
more urgent and better founded each day, that they finally give
the sanitary institutions of the city of Paris the uniformity and
efficacy that they should have long since possessed. (Richet: 1883,
p. 225).

This is the purest expression of the generalized translation referred
to by Serres in his *Parasite*. By taking possession of Pasteurism, which
had taken possession of the microparasites, the hygienists made their
appeals "more urgent" and "better founded." What did they do then?
They *displaced* the engineers. Those jobs are ours! The engineers have
forgotten the microbes in their world and in their plans; we, the
hygienists, make room for them and, thanks to this new authority,
the public authorities must include us in their ranks. Everybody is
displaced, moved, translated. Some lose their places (the engineers,
the microbes, the public authorities): others gain their places (the
Pasteurians, the hygienists). The public authorities are interested in
politics, the engineers in inert bodies, but a new and disturbing agent
has arrived on the scene: living but invisible bodies pullulating everywhere. The Pasteurians say they can see them in their laboratories;
the hygienists believe them. As I have already suggested, politics is
made not with politics but with *something else*. Here was a new source
of power with which to conquer the state. Indeed, the editorial goes
on unambiguously, [the public authorities] "must ensure at last, as
far as possible, the prompt evacuation of refuse, the purity of the
water supply, the cleanliness of dwellings, and the defense of public
health against contagious diseases. It is not only a question of mankind
or public wealth. In a country with such a low birthrate as France,
we must be more careful with human lives; as M. Rochard has just
said, what is at stake is the maintenance of the French nationality"
(Richet: 1883, p. 225). The whole chain has now been described: at
one end, France; at the other, those who in their laboratories make
the microbes visible; in the middle, the hygienists who translate the
data from the laboratories into the precepts of hygiene; a little further
on, the public authorities who legislate on the basis of advice given
by this new profession, scientific hygienism, which must now be taken
into account.

The complete hybridization of hygienists and Pasteurians multiplied the power of both. The least precept in hygiene could now be
dictated by a prestigious, indisputable science, while the most obscure

researchers in laboratories were at grips with the fate of France itself. The archetype of this alliance was not Pasteur but Chamberland, an early collaborator of Pasteur's, a deputy, a hygienist, and the proposer of the 1888 bill on public hygiene. In presenting his celebrated report, Héricourt shows clearly how the secondary mechanism worked. Indeed, for him Chamberland's report reveals public hygiene "as it has been transformed by the researchers of the Pasteur school." He exalts the progress that may be expected: "From the application of all these researchers into microbiology, the initiative for which has come from the laboratory in the Rue d'Ulm" [Pasteur's laboratory at this time]" (1888, p. 245). But this reversal of priorities is not what concerns me here. No, the primary mechanism is more interesting, for it created, using scientific and juridical laws, a new and hitherto unknown public authority. The hygienists wanted to complete the new science very quickly in order to make it indisputable; they now wanted to complete the law in order to make certain obligations irreversible and bring about a change in human behavior. The ratchets of scientific law, juridical law, and public morality must all be turned, one after the other, in order to force the pace of social regeneration and to make room both for the urban masses and for microbes.

Chamberland's report is interesting because it defines explicitly this new authority that was taking nobody's place but was displacing everybody by inventing a new source of "political power." As Rochard writes:

> Already the growing influence of hygiene is offending many a civil servant. "Those doctors are getting everywhere," said a minister a few years ago, somewhat irritated by all the fuss being made about typhoid fever in learned societies and the echoes of their discussions in the nonmedical press. We must expect to be regarded as even more troublesome when the day comes that we shall order instead of advising, when the competent, autonomous authority that we demand will force the municipalities to take the necessary steps and force them to find room in their official expenditure for the sums that such steps require. (1887, p. 388)

A reader of Serres before the event, our Rochard makes full use of his parasitology. There is a lot of talk about typhoid fever; this talk irritates the minister; but the voice does not seem at first very sure of itself; it then becomes a voice that advises; lastly it becomes a voice that orders. It is easy enough to see what this new assurance is based upon: they know what they are talking about; they are talking in the name of bacteriological science, which in turn is talking in the name of that invisible population of microbes which it alone can control. Militant hygiene has begun. It must, our militant continues, "get people's minds used to submitting to the tutelary yoke of this new authority."

Chamberland's report embodies this new voice that has turned up as a third party in all political, economic, and social relations. He proposes in effect "to establish in each department, in the prefect's offices, an authorized agent of public health, who will make sure that the laws are implemented, investigate the salubrity of various communes, and indicate those where work is indispensable" (Héricourt: 1888, p. 248).

A new agent to get rid of the new agents is revealed by microbiology. It's a fine set-up! *For each parasite, a parasite and a half.* Wherever the microbe may find itself, an authorized agent must be there to chase it away. If militant hygiene achieved this aim, it had created a new source of power, a power unthinkable a few decades earlier and one that was rapidly becoming irreversible.

Chapter 2

You Will Be *Pasteurs* of Microbes!

How Are We To Measure the Pasteurians' Displacement?

I have spoken at such length of the hygienist movement in order to reestablish the forces that alone were capable of explaining the immense movement of European society. I had to reestablish, all too briefly, the innumerable crowds and the direction of their general movement in order to deprive the great war leaders, Napoleon or "Pasteur," of the power of performing all these wonders. It would thus be unfair to criticize me for not yet having spoken of the Pasteurians, since I have already described in detail the powers that were attributed to them and on which they capitalized. I have talked as much about them as if, in speaking of an enterprise, I had begun by listing all those who had invested in it, the markets to which it had decided to appeal, and even the several natures of the products that it had decided to manufacture. Pasteurism is *made up* of all this credit. This statement can surprise only those who forget the allies that a science must find in order to become exact. These allies, of which the science is sometimes ashamed, are almost always outside the magic circle by which it later, after its victory, redefines itself. "Pasteur" or "bacteriology" are names given to crowds. Trying to write the history of these phantasmagorias or trying to make one the product of the other would be like writing the history of France on the basis of the popular press filled with crime, sex, or aristocratic weddings.

We must now try to understand what Pasteur, the man—Pasteur without inverted commas—and his team did in this movement. In *War and Peace,* neither Kutuzov nor Napoleon remains inactive, even after Tolstoy has reduced both to the dimensions of men among the crowds that use them and which they in turn use. It is not a question of denying that Pasteur and his team did something.[1] On the contrary, the Pasteurian hagiography is what makes the real work of Pasteur and his followers incomprehensible, since it conceals their own work in a larger whole that includes what others did for them and in their place. Once the process of attributing everything to "Pasteur" has been dismantled, once all the forces offered him have been broken down into their component parts, new questions arise: Did they do anything that was decisive? Did they win the day according to the primary or the secondary mechanism? What precisely did they do on their own? We remember, for example, what Kutuzov, according to Tolstoy, did at Borodino. He had the courage to order nothing himself, to send out again as orders emanating from himself what the commanders had suggested that he do. By a patient study of both *La Revue Scientifique* and the *Annales de l'Institut Pasteur* it is possible to obtain a more precise idea of the work of the Pasteurians.

In order to understand their work, I could have used the word "strategy."[2] But it is not the right word, not because it is pejorative or too political but because it is still too rational to account for the operations in question. As Tolstoy has shown us, the strategists cannot themselves be analyzed in terms of strategy. I cannot therefore analyze scientific credibility by resorting to some other belief: a belief in military leaders. It is enough to speak of "displacement." The Pasteurians *place* themselves in relation to those forces of hygiene that I have described, but do so in a very special way: they go out to meet them, then move in the same direction, then, pretending to direct them, deflect them very slightly by adding an element that is crucial for them, namely the laboratory.

What the Pasteurians did poses no problem to hagiographical history, since it imputes to the ideas of a few men the power of moving everywhere. For a reader of Tolstoy, on the contrary, no diffusion of

a Pasteurian idea, no understanding of a Pasteurian doctrine, no advice, no vaccine, could leave the laboratory without others seizing
upon it, desiring it, having an interest in it. So we must understand
how a few men in their laboratories were accepted and believed. The
first rule of method common to history and the sociology of science
is to convince ourselves that this was not necessary. It might have
been said—it ought to have been said—that this handful of scientists
was precisely no more than a handful. It might—and ought—to have
been said that they were "only theoreticians shut away in their laboratories, without contact with the outside world." This was not said.
Why? If we reject the hagiographical answer, we have to say that they
placed themselves in such a way that the research of their laboratories
would be taken up, as they knew, by people who had been interested
in it.

The control group was provided, even at the time, by the displacement of Claude Bernard. Experimental medicine was already an application of the scientific laboratory to the hospital, but the success
of the Pasteurians, it will be readily admitted, bore no relation to that
of the physiologists, who wanted a strict *separation* between a physiology, proud of its status as an exact science, and a medicine that
was expected to change slowly.[3] There was nothing in common between them and the Pasteurians' takeover bid of medicine, by which
the Pasteurians claimed to be able to "buy," so to speak, the whole
of therapeutics cheaply and to start from scratch again. The laboratory
of Claude Bernard at the Collège de France was in serene and polite
juxtaposition with the art of medicine; that of the Rue d'Ulm claimed
to dictate its solutions directly to pathology. In order to attempt such
an operation without being immediately resisted—and they were not
much resisted—the Pasteurians had to know where to place themselves and to be sure of their allies.

The questions are now becoming clear: How can a laboratory be
made relevant when hygiene and infectious diseases are at stake? How
can the labor power of a few men make all the difference? The general
principle is simple, being the principle of any victory: you must fight
the enemy on the terrain that you master.

The only terrain in which a laboratory scientist is master is that of
experiments, of laboratory logbooks, test tubes, and dogs. This is the
only place where he can convince the adversary, using evidence that
the adversary will not be able to dispute and which will become, as
we say rather thoughtlessly, "indisputable." But the whole problem

Must fight the enemy on the terrain
that you master

[handwritten: 1880 No connection between an infectious disease and a laboratory]

is to carry out a translation, in the terrain of the laboratory, of the enormous problems that are in no way to be found there a priori. We must be careful not to fall into retrospective confusion. In 1871 and even in 1880 there was *no connection* between an infectious disease and a laboratory. To suggest one would have been as odd as to speak in the seventeenth century of a "physics of the heavens" (or to speak nowadays of an anthropology of the sciences). At the time, a disease was something idiosyncratic, which could be understood only on its own ground and in terms of circumstances. This could not be put inside the walls of a laboratory.[4] The hagiographers attribute to the Pasteurians powers that they could not have possessed but omit to credit them with the only things that they did with their little human force. What they did is much more interesting than what they are credited for. Their "contribution," if we insist on this term, is to be found in a certain style of movement that was to allow them to connect "diseases" with the "laboratory." They were to succeed by moving diseases on to the terrain of the laboratory where they, the Pasteurians, had the upper hand. They therefore forced all those groups that were interested in infectious diseases but expected nothing of the laboratory to be interested in their laboratories.[5] In order to succeed in this operation, they had to retranslate what others wanted.

Variation in Virulence

[handwritten: Create interest in Laboratory]

[handwritten margin: Making connections move disease to Lab]

I earlier showed in various ways what the hygienists expected of the new science. I spoke of a fulcrum and showed that the expectations of a science capable of guaranteeing the hygienists' long networks of sanitization over several generations was so great that, if this science had not been offered to them, they would have invented it. Indeed, I showed that they did partly invent it, since they extended and closed it off before it was even operating or even yielding results. Now let us see from the side of those who responded to this request how they transformed the morbid spontaneity of the hygienists into their own terms. Let us take, for instance, the contagions or miasmas of the hygienists. Where can one see them at work? More or less everywhere: in the statistics, in the hospitals, in the nosographical tables, on maps showing the centers of the epidemic. But a Pasteurian would *extract* this contagious ferment and *move* it into an environment that was new and favorable for it, where nothing else would obscure the view of it. This environment was an *ideal* one for the microbe, since for

Laboratory constructed to make invisible agents visible

the first time since the existence of microbes in the world they were allowed to develop *alone*.[6] It was also an "ideal" condition for the observer, since in developing so blithely, the microbe, freed from the competition of other living beings, *made itself visible* by increasing and multiplying. The Pasteurian laboratory was constructed, well before the period under study, in order to make these invisible agents visible.

But a laboratory microbe is not yet a "contagious ferment." It does not have any properties that can retranslate the attributes that were considered as part of the notion of "disease." There may be laboratories of micrography, like those of Miquel, which have no more than a circumstantial relation with hygiene or medicine. A lot of people might be interested in micrographical analyses without yet being able to force the hygienists to go through the laboratory of the Rue d'Ulm. But the post-1871 Pasteur went further. He inoculated animals in his laboratory with the microbe that had been made visible by means of his cultures. He made them ill. He in effect simulated the epidemic. With laboratory-made statistics he counted the sick and the dead and those that underwent spontaneous cure. He performed on dogs, chickens, sheep, what the hygienists did with the help of nationally made statistics on real populations. But because he was operating in a laboratory, the Pasteurian mastered a greater number of elements: the purity of the contagious ferment, the moment of inoculation, and the separation of control groups. What he had created was an "experimental illness," a hybrid that had two parents and was in its very nature made up of the knowledge of the hygienists and the knowledge of the Pasteurians. The double movement of hygienists and laboratory snatched the disease from its own terrain and transplanted it into another. It is easy to understand the growing excitement of all those interested in diseases and the increasing respect with which they treated the laboratory.

Simulated the epidemic

Lab science made all this possible

But the laboratory itself went further. It could have developed an experimental pathology that outlasted the attention and interest of those it had tried to captivate. Instead, it moved one more step *in the same direction* as the interests of the hygienists. By varying the conditions in which the microbes were grown and the conditions of existence of the sick animals, the Pasteurians could now reproduce variation in virulence in the laboratory. That, for the hygienists, was the final takeoff point.

In line with the expectations and demands of the period, the prize

Snatched + Transplanted

Varying Conditions

would go to whoever explained not contagion but *variation* in contagiousness in terms of environmental circumstances. As long as Pasteur could be seen as a contagionist, his laboratory did not have sufficient weight either for the hygienists or for the physicians, since their problem was to reconcile contagions and morbid spontaneity. As soon as Pasteur, using anthrax, reproduced in the laboratory the influence of the environment on the virulence of a microbe, all the power of the hygienist movement shifted and became belief in the laboratory of the Rue d'Ulm. Pasteur was at last doing what was of direct interest to hygienists. He finally synthesized two hitherto antinomic points of view or, in what amounts to the same thing, linked two social groups so that each might strengthen the other. As he himself said: "Work in my laboratory has established that viruses are not morbid entities, that they may affect many different physiological forms and, above all, properties, depending on the environment in which those viruses live and multiply. As a result, even though the virulence belongs to microscopic living species, it is essentially modifiable" (1883, p. 673).

It is scarcely possible to overestimate the importance, at least in the *Revue Scientifique,* of two particular experiments. The reduction in virulence of anthrax cultures by a mere current of oxygen and the triggering of the same disease among chickens, which are not usually subject to anthrax but which contract it when they are placed in cooler temperatures. What struck all the commentators was not the revolutionary character of these experiments but, on the contrary, the fact that all previous hygienists had at last been justified. Duclaux writes:

> I know nothing more striking than that double experiment, which is interesting not only because it holds out the greatest hopes from the therapeutic point of view, but because it brings us the enormous benefit of throwing new light on obscure questions that medicine has hidden behind such terms as receptiveness, organic predisposition, physiological aptitude. In place of these terms and in order to explain the resistance of birds to anthrax, we can state a fact: their temperature is too high and the degree of heat most suited to the globules of their blood is not suited to the bacillus. (1879, p. 631)

The hygiene of the past was both justified and secured on new bases. These experiments were the perfect *exchanger* between the interests of the hygienists and those of the Pasteurians. What took place in the laboratory was what took place in real life: this was the first translation. Variation in virulence was contagion plus the environment: this was the second translation. The consequences were enormous, explaining the whole setup described so far: by pushing Pasteur to his logical conclusions, hygiene both advanced and strengthened itself.

The Contagion Environment or the Traducing Translation

The human or nonhuman agents are interested in some other alliance only if they see that their interests, or what they are led to believe are their interests, are served by it. The alliance of two agents who understand one another very well is to be explained in the same terms as their misunderstandings or disputes. The passion of the hygienists for Pasteur's laboratory is to be explained in the same way as the moderate interest of the doctors in that same laboratory. What the alliances or disputes actually measure is the *angle* of their trajectories. The hygienists accelerate by moving in Pasteur's direction, just as Pasteur's influence grows by responding on his own terrain to others' requests. But this does not mean that the groups understood one another well. Translation is by definition always a misunderstanding, since common interests are in the long term necessarily divergent. Nothing better illustrates this misunderstanding between the agents, even when they get on perfectly well together, than the retranslation by the hygienists of Pasteur's "variation in virulence" (itself a displacement of "morbid spontaneity"). This retranslation bears the name of "contagion environment."

For a French epistemologist used to looking for epistemological breaks, the notion of a contagion environment is an appalling misunderstanding of Pasteur's clear, precise notions. Bouchardat, the "Nestor of French hygiene," as Landouzy calls him, understood perfectly what Pasteur was saying about his experiments on anthrax. He understood so completely that he considered Pasteur to be at last taking Bouchardat's advice seriously: "Morbid ferments, the seeds, if you like, of those diseases, are there permanently and they always find in the Parisian environment terrains that offer favorable conditions" (1883, pp.170–178).[7]

An epistemologist may deride the confusion of the agricultural metaphor. He may say quite rightly that the relation between the seed

epistimologist

and the immune system is quite distinct from Bouchardat's vague notion. He may find it ridiculous to compare Pasteurian medicine with Bouchardat's gibberish, advising in turn vaccination, "the reading of Molière to hypochondriacs," gymnastics, continence. But the historian who is shocked by this mixture will have missed the main point: two distinct generations believed that they understood one another and, acting on this misunderstanding, combined their forces and increased their efficacy.

We must understand this point, which explains both the success of the Pasteurians *and* the continuous choice of their object of research. We cannot at the same time admire the fact that Pasteur was so quickly and so early understood and wish that he had been *properly* understood. By giving impetus to the hygienists' program, Pasteur benefited from the misunderstanding that enabled both groups to declare themselves in agreement. From 1881, in an article specifically dwelling "on the principal modes of attenuating microbes or the morbid ferments of contagious diseases" (1881, p. 458), Bouchardat adopts a protective tone toward Pasteur. Bouchardat can be seen as one of those confused precursors whom the history of science loves to scatter along the way leading to its heroes, but for him the case was almost the reverse. He was the representative of a research program that was determining the way of the Pasteurian hero, who was doing in the laboratory what Bouchardat had wanted to do for a generation. The movement of the Pasteurian research program could be seen as a takeover that, as always, *diverted* the problem toward the place where the Pasteurians knew they were strongest: the laboratory. Nothing could be less revolutionary than this strategy. All the protagonists began to move at precisely the moment when they knew that the old hygiene was vindicated. Again on the decisive experiment involving anthrax, the anonymous author of the *Revue d'Hygiene* writes: "Did not M. Pasteur himself discover the theory of the age-old practice of ventilation by the sanitization of the premises and objects infected? How can he attenuate viruses if he does not subject them methodically, in their culture media, to the action of pure air?" (1883, p. 248). It is Landouzy who invents the perfect hybrid, which he calls "contagion environment." As he says to his students at the beginning of a lecture: "Defining hygiene, the study of men and animals in their relations to their environments with a view to preserving and improving the vitality of the individual and the species, I have chosen as the subject of my lecture the study of the contagion environment. . . This is the

Anthrax to Rabies [handwritten annotation]

environment in which the germs of contagion develop either overtly or covertly, noisily or silently" (1885, p. 101).

The vagueness of this formulation allowed an equivalence of interests to which no sensible man would have given his assent: the macrocosm of the town, sanitized by the hygienists, and the microcosm of the culture of the bacilli, sanitized by the Pasteurians. This truly scandalous short circuit fascinated the *Revue* for a decade or so, from the anthrax period to the rabies period. Public opinion was passionately interested in the esoteric researches of the laboratory in the Rue d'Ulm. All the great macroscopic problems of hygiene, it was believed, had been found to be solvable by the Pasteurians on the small scale of the laboratory: the same went for the main disinfectants, the safety of the Paris drains, the harmlessness of the sewage farm at Gennevilliers, problems of quarantine. In each case, thanks to this identification of the macro- and microcosm, Pasteur's laboratory was expected to provide the final opinion that would settle the matter.

How Pasteur Himself Moved

Cause of a Revolution [handwritten annotation]

Once again, in speaking of the Pasteurians, I have ended up speaking of the hygienists, which is natural enough, since the first had done everything they could to benefit from the strength and knowledge of the second. But how could one man or a few men apply themselves to a whole social movement, then move that alliance in a different direction so that they became, in the eyes of everyone, the cause of a veritable revolution in society? This question, which is usually posed only in politics, must also be posed "in science" as soon as we realize the forces that make up a science. The answer to this question is to be found partly in the period of the journals under study but also partly in Pasteur's career before 1871. What was peculiar to Pasteur was a certain type of movement through the society of his time, a certain type of displacement that enabled him to translate and divert into his movement circles of people and interest that were several times larger. The hagiographers always see in Pasteur's career a necessity, which they therefore omit to admire, whereas they express wonder at the astonishing things that he did not in fact do. A man cannot do a great deal on his own. What he can do, however, is to move. Like the clinamen of the Ancients, this movement uses up little energy but may, if well placed, transform various energies into a vortex that sweeps up everything. This image suits Pasteur perfectly.

Saint story [handwritten annotation]

A man cannot do a great deal on his own [handwritten annotation]

Pasteur began as a crystallographer [handwritten annotation at top]

Pasteur's career has been studied many times. The best analysts, especially Geison, present us with the same enigmas.[8] Certain features of this career have always struck historians as contradictory. Pasteur displays a tenacious obstinacy yet at the same time is constantly changing his object of study; each time he appears as less revolutionary than was said and at the same time delivers a profound shock to the sciences he enters. Dubos, Dagognet, and Duclaux maintain at once that Pasteur pursued a single subject with a single method and was constantly changing the two.

If we agree to simplify somewhat, I may throw some light on all these difficulties by considering Pasteur's sideways movement, only the final sequences of which are found in the period under study. Pasteur began as a crystallographer who interested a dozen or so of his respectable peers and ended up as the deified "Pasteur," the man of a century, the man who gave his name to streets all over France. In fact, what is constant in Pasteur is his movement, regardless of the problems dealt with. Whenever we expect him to pursue the development of a science in which he will have some success, Pasteur chooses not to pursue this fundamental research but to *step sideways* in order to confront some difficult problem that interests more people than the one he had just abandoned. The new problem always appears to be more "applied" than the first one but—and this is the second law of Pasteurian movement—he transforms the "applied" problem into a fundamental problem, which he resolves with the means acquired in the discipline that he has just abandoned. By this peculiar displacement he constitutes each time a new discipline in which he has "some success." He abandons the new discipline in turn in the same sideways movement, and so on. Dubos criticizes Pasteur precisely for interrupting the direction of the fundamental research that he could have carried out. Crystallography, biochemistry, immunology, for instance, are just a few of the disciplines he began and did not continue himself, turning his attention to problems that each time were of concern to a greater number of people.[9]

Pasteur abandoned crystallography but found himself, in the problem of ferments, at the heart of a famous quarrel among the chemists and also at the heart of the beer-, vinegar-, and wine-producing industries, whose economic weight was out of all proportion to that of a few colleagues in crystallography. Yet he did not abandon the laboratory methods acquired in crystallography.[10] Above all, he transformed into a laboratory problem a crucial economic question and

[handwritten margin note, left side: *Turn to problems of That were of a concern to a greater number of people*]

Spontaneous
generation

captured an entire industry that was directly concerned by his experiments. Yet he did not continue his work in micrography, leaving it to others. He moved right into the middle of a quarrel about spontaneous generation. There again he brought onto the laboratory terrain problems that had not previously been there and capitalized on the attention of an educated public that was already much larger than the industrialist public. But he was not interested in developing a fundamental biochemistry. He was put in charge of a new economic problem, that of the silk-worm industry, and there again he transferred all the means of analysis developed in earlier experiments to a new object, disease, which he had not yet confronted. He moved again, and so on, according to a distinctive pattern (see fig. 1, p. 267). Crosses on the horizontal lines show the disciplines that he took over (and was to populate retrospectively with clumsy "precursors"), continuous lines represent the disciplines retranslated by Pasteur, and dotted lines, the disciplines that he left to others to continue. Vertical lines symbolize the sideways steps that took him to a new subject. Concentric circles represent the ever-larger groups that each time he took with him, comprising at first only a few colleagues and becoming in the end what it is no exaggeration to call "the entire world."

As shown in this simplified schema, we can rightly say that Pasteur's career was rectilinear, providing we consider the oblique line that always leads in the same direction. We can also rightly say that he was faithful to a single problem, that of distinguishing the agents involved, for the bent that led him to a new discipline was always the same. Finally, we can rightly say that he was unjust to his "precursors;" he rushed into previous bodies of knowledge with laboratory practices that were different enough to render irrelevant the colleagues who were already engaged in those disciplines. Such a schema also reveals that mixture of audacity and traditionalism found in this strange revolutionary. As Dagognet says, Pasteur innovated by linking together. This ability is not enough for the hagiographer thirsting for genius, but for the historian or sociologist it is essential.

In the period under study this movement of Pasteur became so accelerated and so determined that it eventually took on the regularity of a strategy.[11] Let us look at the speed with which he moved. Scarcely had he made a connection between a contagion and a disease than he stopped in his tracks, leaving others—Koch, for instance—with the job of classifying and describing microbes and their relationship with particular diseases. He set out immediately to find a way of making an experimental disease in the laboratory. But he did not develop an experimental pathology, as perhaps the more prudent Claude Bernard would have done. He immediately looked for a way of attenuating the microbe. Yet microbic physiology as a whole was not what interested him, but the possibility of producing an animal vaccine. As soon as he had this vaccine, he did not confine his attention to experiments which, though interesting, would remain in the laboratory. He immediately set out to extend the methods of his laboratory to the whole of stock-rearing. He could have stayed in veterinary medicine, but this would have gone against the transversal strategy that seemed to become ever more imperious as he reached the end of his course (or what seems to us a century later as the end of his course): to work on the whole of society.

In order to move from the animal to man, from veterinary medicine to human medicine, he chose a disease whose agent, a virus, was to remain invisible until the 1930s and could be cultivated by none of the methods that he himself had developed. Furthermore, after tests on dogs, he passed very quickly to experiments on man. Moreover, he experimented on first one child then two children; he generalized the method, and his next step was toward the vaccinal institute necessary to carry out the research that this general method required and to practice mass inoculation.

As Dagognet rightly insists, none of these stages was a necessary one. To find self-evident the conversion of two cases of cured rabies into millions of gold francs, which were then turned into a laboratory for fundamental research, is not to do justice to the work of Pasteur, the man. All these things were scattered at the time. To link them together would require work and a movement. They were not logically connected. In other words, they did not lay down a particular path. Pasteur could have stopped at any moment and continued himself the work in the fundamental discipline that he was to leave to others. It was even in that direction that all the professional training in the sciences of the time must have urged him. He could have "flinched" at the point where he arrived at human medicine—indeed he did hesitate. He could have—ought to have—not chosen rabies as his first disease; he could have—ought to have—considered the case of Joseph Meister as inadequate to demand the setting up of a research institute. This was certainly what people as different as Peter and Koch criticized him for. Yes, he ought to have done these things, but that type of movement, that audacity, was precisely what defined him, Pasteur—what, indeed, was his *particular* contribution.

It would be pointless to say that there was, on the one hand, Pasteur the man of science, locked away in his laboratory, and on the other, Pasteur the politician, concerned with getting what he had done known. No, there was only one man, Pasteur, whose strategy was itself a work of genius. I am using the word "genius" without contradicting myself, for I am attributing to him nothing that a single man on his own could not do. Let us not forget Tolstoy's lesson. Without any doubt, Napoleon and Kutuzov were at the "head" of their troops. Once the complex of forces that set them in motion is broken down, we have to recognize what those great men did and why Bonaparte and not Stendhal, or Kutuzov and not Miloradovich, entered Moscow. Pasteur placed his weak forces in all the places where immense social movements showed passionate interest in a problem. Each time he followed the demand that those forces were making, but imposed on them a way of formulating that demand to which only he possessed the answer, since it required a man of the laboratory to understand its terms. He began as a crystallographer in Paris and Strasbourg; he ended with "divine honors." Such a metamorphosis does not come about solely by one's own efforts. If he had stayed in Strasbourg, working at crystallography, even his hagiographers have to agree that others would have been accorded the divine honors—even if, as Dubos claims, his researches into the origin of life had been much more important for "pure science." In other words, Pasteur *sought* that glory, and sought it well.

Now that the notion of genius comprises nothing that is not peculiar to Pasteur and is not explicable by displacement and translation, we can understand a little better its most interesting aspect. Pasteur worked just as hard at the primary mechanism, getting allies while he moved, as on what I call the secondary mechanism, getting himself attributed with the origin of the movement. In practice, he always went toward applied subjects that held the interest of a crowd of new people who were not the usual clients of the laboratory; but as he *recruited* his allies in this way, through the needs, desires, and problems that he came in contact with, he maintained a discourse by which all the strength of what he did came from fundamental research and the work of his laboratory. On the one hand, he threw his net as far as he could; on the other, he denied that he had allies and pretended (with the active support of the hygienists and many other groups that needed to take shelter under such a cause in order to advance their own cause more quickly) that everything he did proceeded from "Science." This double strategy bears the stamp of genius, for it amounts to translating the wishes of practically all the social groups of the period, then getting those wishes to emanate from a body of pure research that did not even know it was applicable to or comprehensible by the very groups from which it came. The "application" became a miracle in the religious sense of the term. It was because of this double strategy that the example I have set out to analyze seemed so indisputable. With this double endeavor—recruitment of allies, negation of their efficacy—we end up indeed with the impression that a revolution was emerging from Pasteur's laboratory and spreading into society, which it then turned upside down.[12] The very formulation of what Pasteur did was imposed on his contemporaries (in France at least) by Pasteur himself. I have one more reason for admiring this strategy, which is that a hundred years later it is still at work in more than one philosopher of science. To remain indisputable for so long is surely a lasting victory. Scientific leaders, it must be admitted, are more skillful than military ones. Whereas nobody regards Danton or Lenin as revolutionaries any more, everybody, even in the suburbs, thinks that Archimedes, Galileo, or Einstein carried out "radical revolutions."[13]

The Laboratory as an Indisputable Fulcrum

Having reached this point, we still have explained nothing. Pasteur moved in the way I have described. But nothing proves that in translating into laboratory terms what hitherto bore other names, a person gains enough strength to reinforce both his own position and those of the people who depend on him. In other words, we can wish to do whatever Pasteur wished and still fail miserably. The organization called the Work of Tuberculosis in the same period provides a control group that appeared from time to time in the *Revue*.[14] This association was attacking an infinitely more important disease than rabies but complained, through its founder Verneuil, to be eating up its capital while the Institut Pasteur was being built: "154,000 francs are obviously inadequate and the legitimate agitation about rabies has no doubt done something to make people forget this" (Anon. 1887, p. 444). Like Pasteur, Verneuil was trying to assemble scattered allies who had nothing to make them agree. Neither the success of the one nor the failure of the other was due to the spinelessness of their allies. Verneuil even had in his pocket the so-called Koch bacillus. In spite of this asset, he failed to gather together many interests to struggle against this scourge, which everybody admitted was of the greatest importance. Verneuil's failure reminds us that Pasteur must have done something himself to bring his heterogeneous allies together under his banner. We now come to the heart of the problem, or rather to what has become, by virtue of Pasteur's strategy, the heart of the problem: the microbiology laboratory first of the Ecole Normale and then of the Institut Pasteur.

Methodologically it was crucial for us not to set out from this place. To begin with, the laboratory that was itself the result of a succession of positionings, combinations, and moves would have been sure to give a miraculous vision of its results.[15] We must arrive at the laboratory as the many different actors who found themselves "translated" there arrived. We are now reaching the zone forbidden to sociological explanation, the area that we would like be mysterious, where political and social conditions, which we cannot do other than accept, are transmuted into "truths," "doctrines," and "concepts" that elude all conditions of production and set about, by some miracle that always moves naive souls, to "influence" society.

In fact, I have already indicated the solution, about which there is nothing mysterious. To win, we have only to bring the enemy where we are sure we will be the stronger. A researcher like Pasteur was strongest in the laboratory. Once interests had been aroused in such a way that the macroscopic problems of the hygienists and doctors could be treated at the microscopic level of the laboratory, the procedure was simple enough. A force, even a very small one, applied to the strategic places could bring victory. Everything depended, then, on recognizing those places where this extra force could produce maximum effect. In the laboratory the work of a normal man is scaled up. Pasteur always recognizes this *technical* fact, especially when asking the government for funds:

> As soon as the physicist and chemist leave their laboratories, as soon as the naturalist abandons his travels and collection, they become incapable of the slightest discovery. The boldest conceptions, the most legitimate speculations, take on body and soul only when they are consecrated by observation and experience. Laboratory and discovery are correlative terms. Eliminate the laboratories and the physical sciences will become the image of sterility and death . . . Outside their laboratories, the physicists and chemists are unarmed soldiers in the battlefield (1871, p.1)

Asking for government funds

When it comes to subsidies, Pasteur, as we see very clearly, was as much a materialist as any sociologist of the sciences. The laboratory was the soldier's weapon in the battle. We now know what battle Pasteur was fighting, what strategy he chose to exploit his firepower to the full. Let us now look at the weapon itself.

Why did Pasteur gain strength in the laboratory? He did so because there, as in every laboratory, phenomena are finally made smaller than the group of men who can then dominate them.[16] If this is regarded as simplistic, it is because of not understanding the extent to which the strategy of constructing laboratories obeys this simplification. Roux, bent over a microscope, observing diphtheria bacilli, is stronger than those bacilli, whereas the same microbes, if let loose in nature, laugh at men or kill them. The difference made by the laboratory is small yet crucial. In it the power ratio is reversed; phenomena, whatever their size—infinitely great or infinitely small—are retranslated and simplified in such a way that a group of men can always control them. Whatever the size of the phenomena, they always end up in transcriptions that are easy to read and about which a few individuals who have everything within sight argue. This can be regarded as a miracle of thought, but as far as I am concerned, the simplicity of the procedures by which the balance of forces is reversed is even more extraordinary.

We shall have to understand by what mechanism and skills a handful of men, with nothing but the power of their labor, learned to tame what for thousands of years had secretly frustrated the wishes of all men. This play of minimum and maximum made a great impression on their contemporaries: a microparasite could kill a bull or a man millions of times larger than itself; a few men in their laboratories could acquire in a generation more knowledge about microbes than the whole of mankind from the beginning of time.

Many commentators have insisted on this double disproportion. The infinitely small have been killing us for thousands of years, and the application of a few men was enough to reveal all their tricks: "This microbe of contagion, which we have seized, which we have been able to reproduce through the artifice of its culture in the appropriate liquids . . . it is possible, by exerting upon it certain influences of which the experimenter is master and which he directs as he wishes, to rob it of its excess of energy and to make it, after diminishing its power to the degree necessary, no longer the agent of death, but that of preservation" (Bouley: 1881, p. 547).

men learned to tame microbes

It is not I who am talking about trials of strength but Bouley and, with him, all the scientists of the period.

What Makes Pasteurians Tick

Is it possible to understand the events that took place in the laboratory, which were to have such consequences for all the agents involved? Yes, on condition that we follow the movement of the laboratory techniques as a whole. The contribution of the Pasteurians is easily explained if we follow this movement backward and forward. For convenience, I divide this movement, which I call the "spring" of the Pasteurians, into three stages: in the first stage move the laboratory to the place where the phenomena to be retranslated are found; in the second stage move the phenomena thus transformed into a safe place, that is, where certainty is increased because they are dominated; and in the third stage transform the initial conditions in such a way that the work carried out during the second stage will be applicable there.

This spring of Pasteurism is obviously another way of defining the transversal development but also explains its efficacy. Without it the movement might fail or appear as strategy—perhaps a strategy of genius, but one that would evolve only in the void. Canguilhem attributes this privilege of having a grip on reality to the very nature of the laboratory: "Because the laboratory is a place where the natural data or empirical products of the art are dislocated, a place where the dormant or impeded causalities are freed, in short, a place where artifices intended to make the real manifest are worked out, the science of the laboratory is of itself directly at grips with the technical activity" (1977, p. 73).[17]

To attribute to the laboratory such power is to miss everything that constitutes the spring of the scientific activity. Many laboratories have no grip on anything. Since to understand this spring of action is essential to my purposes, I illustrate it in a simplified way with the example of anthrax.

The first stage is well known. We know how Pasteur or his disciples always visited in person distilleries, breweries, wine-making plants, silkworm rearing houses, farms, Alexandria decimated by cholera, and later, with the Institut Pasteur, all the colonies. Even at the end of our period, during the Great War, Pasteurians still moved their laboratories to the front in order to collect new microbes. This translation of the laboratory is crucial, since it and it alone made possible the capture. The translation always took place on two conditions. On the one hand, the Pasteurians moved but remained men of the laboratory. They brought their own tools, microscopes, sterile utensils, and laboratory logbooks, using them in environments where their use was unknown. On the other hand, they redirected their laboratories to respond to the cause of those they visited.

It was under this double condition that anthrax, a cattle disease, could be redefined as a "disease of the anthrax bacillus." Pasteurians learned from people on the ground—farmers, distillers, veterinary surgeons, physicians, administrators—both the problems to be solved and the symptoms, the rhythm, the progress, the scope of the diseases to be studied. This was the only way of answering all objections concerning the link between the new agent (the microbe) and the old agent (the disease). We should not forget that the bacillus alone might interest a microbiologist but that it was not necessarily the "cause" of anthrax. To get the new agent to do everything that the old disease did, the Pasteurians had to link it, in the most invincibly skeptical minds, with all the symptoms of the disease through spectacular experiments. In order to make this link, Pasteur invented the impossible experiment: he diluted the original bacillus thousands of times, by taking several times a drop of the culture liquid an order to start a new culture, and still caused the complete disease with the last drop of the last culture. He lost his hero on purpose, as Tom Thumb is lost in the fairy story. The bacillus, too, emerged triumphant from the impossible trial. Even when infinitely removed from the animal, the bacillus still causes the disease. It became, therefore, the sole agent of the disease.

The result of these trials was to create a new object that retranslated the disease into the language of the laboratory. Now it was the animal that became like the culture medium: "We inoculate an animal with the bacillus in its pure state; the bacillus develops under the skin as in a culture medium and it gradually spreads" (Pasteur: 1922/1939, VI, 194). Conversely, to grow bacilli in the laboratory is not yet to prove that the soil of Beauce carried them naturally: "These are still only laboratory experiments. We must find out what happens in the countryside itself, with all the changes in humidity and culture" (Pasteur: 1922/1939, VI, 259). This movement from the laboratory to the field cannot be ignored, for by this means new objects intended for the use of future users was formed.

When Pasteur wrote a report entitled "Researches into the Etiology and Prophylaxis of Anthrax in the Department of Eure et Loire: Report to M. Teisserenc de Bort, Minister of Agriculture and Commerce," every word counted: he was addressing a departmental minister. His bacillus would be the cause of anthrax only when it had done everything that the Ministry of Agriculture knew about anthrax. To inoculate the animal with a syringe and give it the disease is all very well, but cows do not get pricked in this way. Pasteur had to invent a way for the bacillus that was credible, so he fed the animals with hay that was infected with cultures. This was not enough. The animals did not die. He then added thistles, thus imitating more and more closely the fields that were known to give the disease. The animal fed with the hay, the bacilli, and the thistles contracted the disease.

We should not underestimate the apprenticeship undergone by the Pasteurians with their predecessors. Their very success, which concealed their role so well, was due to the attention with which they retranslated what those predecessors had said: "I have often heard the knackers, whom I used to warn of the danger that they were running, assure me that the danger had disappeared when the animal was rotten and that one need have no fear unless it was warm. Although taken literally, this assertion is incorrect, it nevertheless betrays the existence of the fact in question (the sporulation of bacteria)" (Pasteur: 1922/1939, VI, 258). What is "betrayed" is rather the transcription of practices by the new practices of bacteriology. The new language can be adopted only if it is made equivalent to *everything* that was said in the old one.

When Pasteurians wrote to a minister, it was not enough to wave a bacillus: they must also be able to say, for instance, what the "accursed fields" mean. For there are many fields that give the disease, even many years later. This is the proof of "morbid spontaneity," to use the language of the veterinary surgeons, or the "curse," in the language of the peasants. To move everyone's belief and replace it by the bacillus, it is not enough to make fun of peasant backwardness. It is necessary to be *stronger* than the accursed field. Koch had already explained the temporal rhythm of anthrax by showing that the microbe could sporulate and survive for years in its dormant form. But Pasteur pursued something that was rather like an ethnographical investigation. He concerned himself with techniques of burying the animals. As the animals lost blood at the moment of burial, they also lost the bacilli, which were sporulating. This explained how the bacilli appeared to "survive." Pasteur now had to explain the appearance of the bacilli on the surface many years later in the accursed fields. "The Académie will be very surprised to hear the explanation for this. Perhaps it will be moved to think that the theory of germs, which has only just emerged from experimental research, has still some unexpected revelations to make to science and its applications . . Earthworms are the messengers of the germ and from its deep burial place bring the terrible parasite back to the surface of the soil" (1922/ 1939, VI, 260). The earthworm! Yet another unexpected new agent to be taken into account. This concern with the field was not the result of friendship for the peasants or of some superfluous amusement. Pasteur knew that only a complete translation would eventually constitute the phenomenon. The "rod bacterium" in the laboratory was incapable of becoming the "cause" of anthrax. It would become so only when Pasteur was able to replace *each* element that composed the definition of the anthrax with his own term and thus convince the minister, the veterinary surgeons, and the peasants, as well as fellow microbiologists.

Through this apprenticeship alone was the microcosm, which seems to reflect the macrocosm so well, gradually built up. And for a very good reason! The Pasteurians constructed the laboratory in such a way as to answer the questions asked of them, but they reformulated them in terms that they understood. Nothing could be more false than to imagine the Pasteurians as overthrowing the old skills with their now clear, distinct methods. On the contrary, they learned those skills but took from them *only* those elements that they could dominate. Who taught and who learned? We do not know. That is why the first stage of the Pasteurian method is also a good translation agency: a new skill emerges from an old skill. We do not have to try to reproduce the whole of epidemiology, but only that which teaches something about the life cycle of microbes; nor the whole of pathology, but only the few symptoms that will enable one to classify the animals infected by the experimental disease. The same process of elimination and structuration that I described in the case of hygiene, to which was added the fulcrum of microbes, is found here at its birth: take a few elements from the field, then reproduce them in the laboratory in new conditions. The crucial element in this extraction and redefinition is to end up explaining with the new actor all the main attributes of the old one.[18]

The Return to the Laboratory

But a Pasteurian does not linger on the terrain of his hastily constructed laboratory. Indeed, the knowledge thus accumulated is almost always weaker at this stage than is that of the men in the field, veterinary surgeons or physicians. The whole Pasteurian strategy, now that they have extracted a few aspects from the macrocosm, is to gain strength by making a long detour to their central well-equipped laboratory. This is the second stage. Geographically this stage usually takes the form of a return to Paris or a veritable shuttle diplomacy between Paris and the provinces. It is here, of course, that the redefinition by Pasteurians of the questions posed by what is now "outside" become indispensable.

Should we now suspend our analysis in terms of trials of strength from the moment when we have at last reached the microbe "discovered" by Pasteur? Have we arrived at places, methods, types of agents that differ from those we have so far studied? Do we notice, as we retrace the Pasteurian path, that we have crossed a sacred fence? No, of course not, for urban microbes are made of the same stock as country microbes. We do not know beforehand what an agent is doing. We must try it out. This one corrupts a veal stock; that one transforms sugar into alcohol; the other one survives in gelatine but is interrupted in urine. How are we to define a shape? Like all the others: they are the edge of trials of strength that others subject them to. If we boil water five degrees more, a new species is then defined, whose "edge" is to resist the temperature of 100°. If we deprive it of oxygen, then others are defined that do not need air.

Since microbes saw their forms stabilized before the period under study, it is difficult to recall the time when they were being forged and tested, like Siegfried's sword.[19] But take, for instance, this new agent that appears on the scene, in the 1890s, which is defined by the list of actions it made, and which as yet has no name: "From the liquid produced by macerating malt, Payen and Persoz are learning to extract through the action of alcohol, a solid, white, amorphous, neutral, more or less tasteless substance that is insoluble in alcohol, soluble in water and weak alcohol, and which cannot be precipitated by sublead acetate. Warmed from 65° to 75° with starch in the presence of water, it separates off a soluble substance, which is dextrin."

The Greek name should not make us forget the tests, for it is the name of an action, like Indian names. Instead of He-Who-Fights-the-Lynx, we have He-Who-Separates-Starch. The object has no other edges, apart from these tests. The proof is that we only have to change these tests to define a new agent: "A more extended contact of the diastasis with the starch paste in turn converts dextrin into a sugar, which differs from dextrin in that it is no longer precipitated by sublead acetate" (Duclaux, 1898, p. 8).

In the laboratory any new object is at first defined by inscribing in the laboratory notebook a long list of what the agent does and does not do. This definition of the agent is acceptable, but it runs the risk of bringing us a new philosophical problem. Did the microbe exist before Pasteur? From the practical point of view—I say practical, not theoretical—it did not. To be sure, Pasteur did not invent the microbe out of thin air. But he shaped it by displacing the edges of several other previous agents and moving them to the laboratory in such a way that they became unrecognizable. This point is not unimportant, for we often say without thinking that Pasteur "discovered" the microbes. Let us see some of these displacements which practically solve the problem that realist philosophers often have with the history of science.

The first anthrax, as I have said, had previously been defined as a disease. Its edges had been defined by cows, wounds, corpses, accursed fields, veterinary surgeons, and Rosette, who had such a beautiful hide. The earlier application of a science, which had also come from Paris or Lyon, had already altered that disease and turned it into an epizootic disease whose edges, this time, were a set of patches on the map of France, where we could count its sites, follow its wanderings, and detect its recurrences. The agent constituted by epidemiology was rightly called "the anthrax epidemic," in order to sum up all the statistical trials that defined it. Predecessors—those who became predecessors like Arloing or Davaine—had already brought their laboratory into contact with anthrax. But the new actors did not become more visible for all that. The tests did not convince the observers. The link between anthrax and a contagion remained debatable; that is, more or less anyone could, without great effort, make several statements on the subject that were just as plausible as those by Davaine or Arloing. In going to the laboratory, these authors did not put an end to the controversy but increased it. This happened because their laboratory, which had already become an obligatory point of passage, was not capable of translating into its own language all the phenomena associated with anthrax. Bypassing it was still easy.

To discover the microbe is not a matter of revealing at last the "true agent" *under* all the other, now "false" ones. In order to discover the "true" agent, it is necessary in addition to show that the new translation also includes all the manifestations of the earlier agents and to put an end to the argument of those who want to find it other names. It is not enough to say simply to the Académie, "Here's a new agent." It must be said throughout France, in the court as well as in town and country, "Ah, so that was what was happening *under* the vague name of anthrax!" Then, and only then, bypassing the laboratory becomes impossible. To discover is not to lift the veil. It is to construct, to relate, and then to "place under."[20]

In this transformation of the agents, everything depends on the new trials. Place a sterile pipette on Rosette's wound, take blood, place a few drops of it in urine—these are the new *gestures*. The translation of the agents is not intellectual or linguistic; it is found entirely in the skill. Taking blood is no more abstract, more rational, more rigorous, more ideal, than milking a cow. Moving from the farm to the laboratory, we do not move from the social to the scientific or from the material to the intellectual. The difference comes from the fact that the world of the pipette, the culture medium, and the guinea pig is a world-to-grow-the-microbe, just as that of the farm is a world-to-rear-cows. Indeed, the laboratory itself is constructed only out of the movement and displacement of other places and skills. The culture medium, for instance, is at the beginning very close to a cooking stock. It is not transubstantiated when Duclaux manipulates it: "One obtains a culture medium by leaving for twenty-four hours, in contact with twice its weight in water, finely chopped lean veal. One strains off the liquid, presses out the residue, cooks the resulting liquid for an hour and strains it. One then adds 1% peptone and 0.5% sea salt and enough sodium solution to make what is usually a slightly acid liquid neutral" (Duclaux, 1898, p. 105). To make a gelatine, "one adds a white of egg, extended by five or six times its weight in water." These details are not ridiculous. They are the body and soul of the things we are discussing, as Pasteur himself said. Nothing could be more wrong than to imagine that the farm is to the laboratory as the first degree of reflection is to the second degree, as practice is to theory, as "praxis" is to "knowledge." The laboratory is to the farm what Duclaux's medium is to soup.

But to understand more clearly the relation between the Pasteurians and the microbes that they revealed in the laboratory, we must stress the fact that, although the trial is new for the Pasteurian, who has never yet had to take a microbe from a cow, it is even more so for what will become the microbe. Or rather, the creation of culture media is just as much a historical event for the microbe as for the Pasteurians. There is a history of microbes that is also filled with sound and fury.[21] History is no more limited to the so-called human agents than to the nonhuman agents. What were once miasmas, contagions, epidemic centers, spontaneous diseases, pathogenic terrains, by a series of new tests, were to become visible and vulnerable microorganisms. Why? Because for the first time in the history of the world (a solemn tone is not out of place here), the researchers of the Rue d'Ulm were to offer these still ill-defined agents an environment entirely adapted to *their* wishes: "Urine is an excellent culture medium for the bacillus; if the urine is pure and the bacillus pure, the latter will multiply promptly" (Pasteur: 1922/1939, VI, 199). For the first time these agents were to be separated out from the confusion of competitors, enemies, and parasites, which hitherto they had to take into account. For the first time—for them as well as for us—they were to form homogeneous aggregates. This was the decisive advantage of the solid media later invented by Koch: "The gelatine medium forces each germ to develop on the spot and to form a *colony,* which soon becomes visible to the naked eye and whose form, color, growth, superficial or profound, and action on the gelatine, are so many characteristics ready to be consulted, some of which even, in given circumstances, may become characteristic" (Duclaux: 1898, p. 104).

Isolated from all the others, microbes grow enthusiastically in these media, which none of their ancestors ever knew.[22] They grow so quickly that they *become visible* to the eye of an agent who has them trapped there. Yes, a colored halo appears in the cultures. This time it is the man bent over the microscope who is enthusiastic. This event completely modifies both the agent, which has become a microbe, and the position of the skillful strategist who has captured it in the gelatine. Without this transformation's being made on the microbes, the Pasteurian would have been without a fulcrum. He was now going to be able to modify the culture medium, starve the microbes, kill them with antiseptics, make them eat anything, in short, torture them in innumerable ways, in order to learn something about them each time.

What does "to learn" mean in such a context? Are we to arrive at last at that mysterious world of ideas which seems to float over the colonies of microbes and to enable us to escape from all trials of strength. Have we passed the line? No, for to learn is simple enough. It means to note the culturings, number the Petri dishes, record times, look things up in the archives, transfer from one page to the other of laboratory logbooks the answers given by the tortured or, if a less harsh word is preferred, "tested" or, an even gentler word, "experimented on" objects. In *inscribing* the answers in homogeneous terms, alphabets, and numbers, we would benefit from the essential technical advantage of the laboratory: we would be able to see at a glance a large number of tests written in the same language. We would be able to show them to colleagues at once. If they still disputed our findings, we would get them to examine the curves and dots and ask them: Can you see a dot? Can you see a red stain? Can you see a spot? They would be forced to say yes, or abandon the profession, or in the end be locked up in an asylum. They would be *forced* to accept the argument, except to produce other traces that were as simple to read—no, *even simpler* to read.

Although the laboratory is constituted only by displacement and transfer, it makes an enormous difference in the end. On the farm there are calves, cows, clutches of eggs, Perette and her milk jug and the willows beside the pond. It is difficult to locate Rosette's disease or to compare it with another. It is difficult to see anything at all if what we are looking for is a microbe. So we are doomed to argue endlessly about the disease. In the laboratory, the researchers have colony no. 5, no. 7, no. 8, with control colonies no. 12, no. 13, no. 15. A double-entry system with crosses and spots. That is all. We have only to be able to read records. The argument (if it is about these spots) will end. A lot of things may be learned on the farm, but not how to define microbes, which can be learned in the laboratory. The issue is not that the first has an ontological superiority over the second; it is simply that the laboratory draws on everything—not milk, eggs, firewood, and the hand of the farmer's daughter, but sheets of paper that can be easily moved and placed on top of one another and can be argued about at leisure as if we were "on top of the question."[23]

In the laboratory unprecedented things were now to be expressed in written signs. Impossible superimpositions were to take place, movements that would have required considerable energies took place, from sheet of paper to sheet of paper, over a few centimeters. For instance, it would be possible to compare the written form of all those newly defined microbes. Where could we do such a thing? Only in the laboratory, once the microbes had all been written down.

It is less surprising if, in the middle of all these accumulated traces, even error becomes useful. That Chamberland forgot a culture, that Pasteur proposed to inoculate animals with it, and that these animals survived longer after a new inoculation of fresh bacteria are the types of event that could happen only in a laboratory. The attenuation becomes visible only in the middle of well-recorded control groups. Even the aging of the culture is detectable only on the pages of a well-kept laboratory logbook. The event, though unexpected, is normal, since it is for just such an event that a laboratory is constructed. As Pasteur might have said: "Chance favors only well-prepared laboratories." If we make new actors simultaneously visible, we see new things. We must have our faith well secured to our body to find in this tautology a mystery that would separate "the science" from everything else.

In other words, the laboratory, directed entirely toward a reversal of the balance of power, also has a history. The unexpected opportunity—a forgotten culture—immediately becomes a method. Laboratories convert chance to necessity: "The method of preparing the attenuated virus is wonderfully simple, since one has only to cultivate the very virulent bacillus in chicken stock at 42-43° and leave the culture after its completion in contact with air at this same temperature" (Pasteur: 1922/1939, VI, 343). What we have to admire is not this false mystery that claims to elude trials of strength under the pretext that these people are wearing white coats, but the cleverness of this *reversal* of the balance of power. The microbe itself, somewhat weakened, serves as a double agent and, by warning in advance the immunitary field, betrays its companions. In order not to see in vaccination a story of strength and weakness, we must again have great faith, a faith that resists all questions.

But I have not answered my own question. Did the anthrax microbe exist before Pasteur? We do not know yet. We always state retrospectively the previous existence of something, which is then said to have been "discovered." In order to separate invention from discovery, the product of the laboratory from the "fact of nature," we need a little more. We need Pasteur and his colleagues successfully to complete the third stage of their movement. The single term "anthrax bacillus" must be made to serve as a translation for everything that used to be covered by the term "anthrax." Without this link and translation, Pasteur would have had a microbe that performed certain things *in* the laboratory and a disease left to itself *outside* the laboratory, with endless talk filling the gap. He would not be said by others to have "discovered" the cause of anthrax. In order to produce the retrospective impression that the anthrax bacillus had been there from the beginning of time and had been covertly active before Pasteur surprised it, Pasteur had to link each gesture in the laboratory with each event associated with anthrax on the farm by extending the trials that formed the microbe. To do this, the laboratory had to be moved again so that it was actually in contact with each trial and could retranslate it into its own terms. In order to carry off this new coup, Pasteur had to have more than one trick up his sleeve and keep up more than one network.[24]

The Theater of the Proof, or How To Become Indisputable by the Greatest Number

To go from laboratory trials back to life-sized tests, the trials must themselves be life-size. It was on this condition that not only the few colleagues and collaborators but also all those who needed to understand anthrax would accept as indisputable the redefinition of anthrax as "the disease caused by the anthrax bacillus." If Pasteur's experiments were so demonstrative, as we say, it is because they were invented with the aim of definitively convincing those whose interest had been moved on the return from the field to the central laboratory. Pasteur's genius was in what might be called the *theater of the proof.* Having captured the attention of others on the only place where he knew that he was the strongest, Pasteur invented such dramatized experiments that the spectators could see the phenomena he was describing in black and white. Nobody really knew what an epidemic was; to acquire such knowledge required a difficult statistical knowledge and long experience. But the differential death that struck a crowd of chickens in the laboratory was something that could be seen "as in broad daylight." Nobody knew what spontaneous generation was; it had given rise to a highly confusing debate. But an elegant, open, swan-necked bottle, whose contents had remained unalterable until the instant the neck was broken, was something spectacular and "indisputable." It is important to understand this point, for the hagiographers had a tendency to separate what Pasteur's genius brought together. He had to perform such telling experiments because he wanted to convince the outside forces that he had recruited at the outset. As Bouley says: "He is not one of those whose virtue remains idle when they have to make their opinion prevail" (1881, p. 549). To say the least! Pasteur did not wait for his ideas to emerge fully formed from his laboratory and allow them to spread through society. He gave them a lot of help. The greater the groups that he wanted to convince after capturing their attention, the harder he hit. All commentators agree about his violence in argument. Even the hagiographers use phrases to describe Pasteur's rhetorical activity that might be better suited to the much-despised politicians: "Master of what he knew to be the truth, he wanted—he knew—how to impose it by the evident clarity of his experimental demonstrations and to force most of those who had proved to be most unwilling to do so at first to share it with him" (Bouley, 1881, p. 549).

To "force" someone to "share" one's point of view, one must indeed invent a new theater of truth. The clarity of Pasteur's expositions is not what explains his popular success; on the contrary, his movement to recruit the greatest possible number of allies explains the choice of his demonstrations and the *visual* quality of his experiments. "In the last instance," as one used to say, the simplicity of the perceptual judgment on which the setting up of the proof culminated is what made the difference and carried conviction. Pasteur was not stinting in the laboratory and outside in concentrating interest and discussion on a few extremely simple perceptual contrasts: absence/presence; before/after; living/dead; pure/impure.

Of course, the laboratory also accumulates a large number of trials and data that remain unknown to the public. Pasteur is nevertheless to my knowledge the only researcher who was able to interest a large educated public in the well-nigh daily drama of his experiments. Let us not forget that for almost a decade the *Revue Scientifique* followed week by week the research being carried out in the Rue d'Ulm. This situation was a long way from that of the laboratory isolated from society whose benefits would later take the form of technical results. No, the very genesis of the data was followed step by step. I have shown why there was this passion on the part of the hygienists—the contagion environment made it possible to identify macroscopic hygiene and the laboratory—but I must now show how interest was maintained on the part of the Pasteurians by ever more astonishing experiments. The effect was all the more powerful in that the laboratory, on the basis of its own problems and procedures, was producing results that each time justified, simplified, or strengthened the task of the hygienists. Without this double movement of interest and dramatization, the tests might have been indisputable but would either have remained undiscussed because they interested nobody or, like those of Davaine, have been discussed but remained disputable. With his capture of interests, on the one hand, and his theater of proof, on the other, it would be unfair not to grant him genius.[25]

Pouilly-le-Fort

Although the form assumed by the skills in the Pasteurian laboratory was not exceptional, it was specified. Indeed, even if the Pasteurians developed a biology *in* the laboratory, they did not practice a *laboratory* biology. They did not leave to others, as apparently happened in England, the job of using or applying their results, contenting themselves with "pure science."[26] No, they went on winding up a dramatic plot and moving the laboratory once again, but this time in the opposite direction. In this third stage they set out to transform the field from which they had come according to laboratory specifications, in such a way as to retain the balance of power that they had been able to reverse in the second stage. The third stage is the most spectacular, and "the wonderful experiment of Pouilly le Fort" is the archetype of it in the pages of the *Revue Scientifique*. As Bouley writes: "Pouilly-le-Fort, as famous today as all the battlefields . . . Monsieur Pasteur, a new Apollo, was not afraid to deliver oracles, more certain of success than the son of poetry would be" (1883, p. 439).

What has not been written about Pouilly-le-Fort! Yet it was only the final episode in this theater of proof; the dress rehearsals were over. The Pasteurians were now playing life-size before the assembled media. Pasteur predicted that a lot of unvaccinated sheep would die: "The same number of animals, which had been covered with the palladium of the new vaccine, remained invulnerable to fatal inoculation, and were shown, very much alive, surrounded by corpses" (Bouley: 1881, p. 548). There was talk of miracles, but we have to understand what gives the *impression* of a miracle. If we forget the other two stages in the Pasteurian strategy, then indeed his prediction becomes extraordinary, wonderful, truly divine. This miracle belongs with the one that many philosophers have tirelessly celebrated, the *adequatio rei et intellectus*. The predictions of the laboratory have an application in reality. Yet in practice, the prediction is at once less extraordinary and more interesting. Even Bouley, the chief thurifer, is obliged to admit as much: "In a program laid out in advance, everything that was to happen was announced with a confidence that simply looked like audacity, for here the oracle was rendered by science itself, that is to say, it was the expression of a long series of experiments, of which the unvarying constancy of the results proved with absolute certainty the truth of the law discovered" (1881, p. 548). The impression of a miracle is provided by the great break between the "laboratory," in which scientific facts were made, and the "outside," where these facts were verified. The impression disappears, according to Bouley, if one considers the long, continuous sequence of experiments.

But we cannot have it both ways: either the certainty was absolute, in which case there is nothing to get excited about, or it was not, in which case it was a gamble that might have gone wrong. Pasteur, as always, says much more about this than his admirers. In a famous passage that is rarely quoted in its entirety, Pasteur constructs a rule of method, written in the style of Napoleon at Arcole: "This program, I admit, had the boldness of prophecies that only a dazzling success could excuse. Several individuals were kind enough to point this out to me, adding a few criticisms concerning my scientific imprudence. However, the Académie must understand that we have not drawn up such a program without a solid basis in earlier experiments, though none of those experiments had the scope of the one we were planning. Indeed, chance favors prepared minds, and it is in this way, I believe, that one must understand the inspired words of the poet: *audentes fortuna juvat*" (1922/1939, VI, 348). The scope that was to be given to the experiment was not exactly a guarantee. Conversely, the experiment was not entirely without guarantee. There was a risk. This risk was contrary to scientific prudence. But Pasteur was well aware that the epistemology of falsification was false: that only the victor is reasonable is the essential principle. A "dazzling success" may alone justify the risky crossing of the bridge at Arcole. The audacity that is only an "appearance," according to Bouley, is on the contrary real, according to Pasteur. His magnificent strategy explains the spring that he was trying to wind up: set out from the farm and return to it a victor, but without being too certain of it, expecting Fortune to do the rest, that extra bit which could bring the little laboratory to the life-sized farm. We know that he was nervous; we can see why (E. Duclaux, 1896/1920; L. Nicol, 1974).

Let us follow the cavalcade again. The laboratory first moved to a farm in order to capture the bacillus (first stage). Then in the central laboratory the bacilli cultures were moved, purified, and inoculated on an entire experimental farmyard; there, too, the bacterium was weakened and the animals' bodies were strengthened by vaccination (second stage). Lastly, after the transformation of a farm in such a way that it partly obeyed the conditions of a laboratory experiment and maintained the reversal of the balance of power, the experiments carried out in the central laboratory were repeated (third stage). The whole of the Pasteurian arc is indeed dazzling, but the experiment of Pouilly-le-Fort in the context of the arc is not miraculous. It is better than that. It has cheek.

In the course of this third stage is found the same phenomenon of negotiation as in the first. Indeed, the account of the experiment discussed with the Société d'Agriculture at Melun is called a "convention program." Just as some elements of the field were taken to the laboratory, so certain elements of the laboratory were taken to the field. It was absolutely impossible to show the Société the efficacity of this vaccine if the farm was not to some extent transformed into a laboratory annex. For instance, the vaccinated animals had to be separated from the nonvaccinated ones and marked by a hole in the ear; every day their temperatures had to be checked and recorded; syringes had to be sterilized; and more and more control groups had to be supplied.[27] None of this happens on an ordinary farm. The farm known to Perette and Rosette had in turn to be moved so that the defeated microbes could this time become visible there. For all this, negotiations had to be carried out with the organizers so that the results of the laboratory could be transferred. The negotiations were delicate, for if Pasteur imposed too many conditions, his vaccine would remain in the laboratory and become immovable; on the other hand, if he departed too much from "previous experiments," the effect of his vaccine might run the risk of no longer being detectable and the whole thing could turn into a fiasco.

Once Fortune had smiled on the brave man who had added to scientific prudence a life-sized experiment the results of which could not be guaranteed, Pasteur applied this double strategy. On the one hand, he constructed the whole of the arc that would enable him to create, strengthen, and extend the work of the laboratory (the primary mechanism); but on the other hand, he imputed or left others to impute all the predictions to "Science" (the secondary mechanism). That the truth of the laboratory could be applied at Pouilly-le-Fort then became a miracle, accepted by all as such, of which he became the prophet. This double movement was admirable but not magical. For me, the most incomprehensible thing about the universe is that anyone should regard as incomprehensible the nevertheless simple way by which we make it comprehensible.

From the Micro- to the Macrocosm

All the Pasteurian "applications" were "diffused," as we say, only if it was previously possible to create *in situ* the conditions of a laboratory. The pasteurization of beer or milk, hermetically concealed containers, filters, vaccines, serums, diagnostic kits—all these served as proof, were demonstrative and efficacious, only in the laboratory. If these applications were to spread, the operating room, the hospital, the physician's office, the wine grower's winery, had to be endowed with a laboratory. Of course, the entire laboratory in the Rue d'Ulm did not have to be moved or reproduced, just *certain* of its elements, certain gestures, certain procedures, which were practiced only there and were indispensable to maintaining in existence the phenemenon in question. Indeed it was by holding to this last stage of the work that the Pasteur Institute and its subbranches were to have a lasting means of occupying the field.

If Pasteur had written a work on the sociology of the sciences, he might have entitled it "Give me a laboratory and I shall raise the world" (Latour: 1983b). As with Archimedes' lever, the fulcrum of the laboratory on domesticated microbes made possible a *real displacement* of the world. The case of anthrax shows why: Pouilly-le-Fort might have remained an isolated case, an interesting prototype with no future. But Pasteur did not wait for the future; he recruited it. Pouilly-le-Fort was a large-scale theater within which with overwhelming arguments he could convince equally enormous social groups—the French stock-rearing industry, the Société d'Agriculture, government ministers. Indeed, the third stage was not yet complete. On June 1, 1882, the laboratory's field of action was geographically extended to the extent that 300,000 animals, including 25,000 cows, were vaccinated. According to the statistics, the mortality rate fell from 9 percent to 0.65 percent. The editorial continues: "Confronted by such figures, we can no longer doubt the efficacy of the vaccination against anthrax" (Anon.: 1882, p.801).

So we have returned to the macrocosm, to statistical data, to the old epidemiology, to precisely that from which we set out and which showed on the maps, the erratic expansion of the centers of anthrax infection. Now the same statistical apparatus was used by the Pasteurians as a great account book to follow, life-size, that vast laboratory experiment on the scale of France itself, by which the rate of the death of vaccinated animals was compared with that of nonvaccinated ones.

Have we returned exactly to our starting point? No, for there was another movement. In the tables there are now two parallel columns: before vaccination and after vaccination (Pasteur: 1922/1939, VI, 383). This is why I spoke of the lever. Phials of vaccine produced in Pasteur's laboratory were distributed under supervision to all farms and displaced the anthrax bacillus. The wild microbe would no longer act as a parasite on sheep; thus, we, the men who are the parasites of sheep, could use the domesticated microbe, that is, the attenuated bacillus, to save our domesticated animals, which, in getting fat, fatten the farmer and us. Everyone along the parasitic chain gains, except the anthrax.

Had we not followed simultaneously Pasteur and his allies, there would have been, on the one hand, a science and, on the other, a society. The correspondence between the two would have then become a miracle. If we follow all of them, we are able to see a continuity in the trials of strength based at first on the home ground, then moved to the laboratory, then reversed, then used to move the terrain from which they started. The result of such an arrangement, such a twist, is that a minimum of effort—a man, a few bacteria, a few years of work—has in the end the maximum effect. The interest in this way of seeing things is that we avoid the error with which we set out. Pasteur's work does not "emerge in society" to "influence" it. It was already in society; it never ceased to be so. The very existence of anthrax as an agent disturbing the peace of the countryside depended on a first science, statistical epidemiology. This anthrax was no more "outside" than Pasteur's anthrax. It was simply in the offices of the Ministry of Agriculture, obtained by movements of civil servants, researchers, and inspectors, which made it possible to obtain the mortality figures and, in a single spot, the statistics. This evidence is always forgotten: neither the existence of anthrax as a national danger nor the efficacy of Pasteur's vaccine as a national salvation would have been visible without this first *measuring apparatus,* the statistics of the Ministry of Agriculture, whose history must be written in the same terms.

The same goes for the "efficacy" of Pasteurism as for the "discovery" of the microbes or for the "predictions" of Pouilly-le-Fort. They were miracles if we forget that the Pasteurians created in advance networks in which they could reverse the balance of forces. It is a general rule that we never observe in "science" or "technology" the emergence of a result, a process, a technique, or a machine. The "outside" of such networks would be so unpredictable, so unorganized, that the laboratory results would be once again inconsequent. They would have the lower hand; they would become disputable.

If the reader hesitates for a moment before being convinced of this network quality of science, he has only to look at the polemics in which Pasteur engages on the subject of extending the application of the anthrax vaccine. There again Pasteur knows very well that the vaccine is not a product that travels alone. It must be followed closely. In the sessions of the Académie, Colin is Pasteur's whipping boy, for he doubts the efficacy of the anthrax cultures. Pasteur replies that his doubt depends on his *hands*:

> *Colin*: This liquid was active at the beginning; I have watched it kill rabbits and even a sheep; but after a few months it no longer brought on anthrax. Why it became inert I don't know.
> *Pasteur*: Because it had become impure in your hands.
> *Colin*: But it had never been in contact with the air.
> *Pasteur*: How did you uncork it?
> *Colin*: I took all the precautions indicated by Monsieur Pasteur and his assistants to draw it out of the apparatus and to reclose it . . .
> *Pasteur*: In fact, allow me to say that you need to study these questions a good deal more . . .
> *Colin*: [The germ corpuscles] were to be found there in large numbers and with their original characteristics.
> *Pasteur*: I have already told you that that is impossible . . . But you do not know how to withdraw this organism in such a way as not to introduce a new impurity into the tube. How do you expect us to give any credence to your assertions? You are still seeking not the truth but contradiction. (Pasteur: 1922/1939, VI, 357)

Everything is there: "How did you uncork it?" Colin is the idealist. He is the one who believes that there are ideas in science. But no, one gains credit with one's fingers. Phenomena remain only as long as one maintains them within the trials of strength. A bacillus is present only as long as a set of gestures that guarantee its purity, just as a telephone message is maintained only along a line. "Ideas" never escape from the networks that make them. It often seemed for instance, that the antianthrax vaccine refused to pass the Franco-Italian border.[28] However much it tried to be "universal," it remained local. Pasteur had to insist that the practices of his laboratory be repeated exactly if the vaccine were to travel.

Techniques are neither verified nor applied; they can only be repeated and extended on condition that their movement is prepared by erecting the whole apparatus that enabled the phenomena to exist. There is nothing extraordinary in this. A theater of proof, like the ordinary theater, needs all its accessories. The shape of the microbe is only the relatively stable front of the trials to which it is subjected. If we stop the culture, if we sterilize the pipettes badly, if the incubator varies in temperature, the phenomena disappear; that is, they change their definitions. From this point of view the Pasteurians are like the hygienists. They need the safety of long networks for the "truth" that they declare to be made indisputable on all points.

"Well, Mister Know-It-All, did Pasteur discover the cause of anthrax or not?"

Now I should like to reply at last in the affirmative. But this affirmative is also accompanied by a lot of accessories. Once the statistical apparatus that reveals the danger of anthrax and the efficacy of the vaccine, has been stabilized, once at the Institut Pasteur the procedures for weakening, conditioning, and sending the vaccine microbe have been stabilized, once Pasteur has linked his bacillus with each of the movements made by the "anthrax," then and only then is the double impression made: the microbe has been discovered and the vaccine is distributed everywhere.[29] This double projection in time and space is not false; it only takes long, like any projection in the cinema, to construct, to focus, and to tune. I would be prepared to say that Pasteur had "really discovered" the truth of the microbe at last, if the word "true" would add more than confusion.

The Style Is the Pasteurian

Having described Pasteur's sideways movement and the very special strength that enabled him to hold together what seemed disjointed, I will become a little more ambitious and define with some precision what was meant to be a Pasteurian in this period. I grasp it as a term of *style*. Just as there was a hygienist style, there was a style that could be recognized at first glance as Pasteurian. A scientific article is not of course a description or a distraction. It is a means of pressure on readers, convincing them to change what they believe, what they want to do, or what they want to be. In order to construct those paths that attract, move, force, or do violence to the reader, the author associates himself with everything that may tend to support his point of view and to make his conclusions as indisputable as the course of a river through a V-shaped valley. This much is general. The particular feature of the Pasteurian articles is that they orient the reader along a slope, but this slope encounters orders of preoccupation absolutely alien to one another. This characteristic enables the Pasteurians to "tie" together the forces that they capture, to use them, and thus to increase the weak forces that they throw into the battles of the time. To take another metaphor, the Pasteurians cross *diagonally* the front of the adverse forces and, by this movement, become their leaders, deflecting their movement through the help they find in each of the other groups. The subtlety and elegance of this style of action may be shown, astonishingly enough, in a single article as well as in the total corpus of the *Annales de l'Institut Pasteur*. To begin with, let us take an article written by Yersin, entitled "The Bubonic Plague at Hong Kong" (1894, p. 663).[30]

Yersin begins by summarizing the contribution of a century-old science, epidemiology, one of whose aims was to throw back the microscopic "barbarians" beyond the frontiers of the western world: "At the beginning of last May, there broke out at Hong Kong an epidemic of bubonic plague that proved deadly to the Chinese population of that city. The disease had been raging for a very long time in an endemic state on the high plateaus of Yunnan and from time to time had appeared quite near the frontier of our Indo-Chinese possessions at Meng-tsu, Lan-Chow, and Pei-hai." Since the time of the Renaissance, the contradiction has always been the same: to extend commercial routes was to allow microbes to multiply. "Laissez faire, laissez passer" profited not only the merchants: "The great commercial movement between Canton and Hong Kong, on the one hand, and between Hong Kong and Tonkin, on the other, and the difficulty in establishing on the littoral of these lands a really effective quarantine makes the French government fear that Indochina will be invaded by the epidemic."

There is still nothing original about the beginning of this article, which conforms to all the canons of hygiene. The state defends its frontiers with soldiers against large-scale enemies and with doctors against small-scale ones. In Tonkin the French go only as far as they are allowed to by the pullulation of parasites and microbes that secretly undermine not only the colonial officers but also the Tonkinese and, of course, their dogs, cats, and buffaloes: "I received from the Ministry of the Colonies the order to go to Hong Kong and to study the nature of the scourge, the conditions in which it spreads, and to seek the most effective measures to prevent it reaching our possessions." The plague could sweep away "possessions," just as anthrax could decimate French livestock. There, too, a minister entrusted a Pasteurian with a mission. There was nothing original in that decision. The century saw innumerable great missions of inquiry into ways of protecting parasites (white-skinned macroparasites) against parasites (microparasites in the form of miasmas or centers of infection). At this point, however, the article takes a quite different turn: "I set up my laboratory in a straw hut that I had built, with the permission of the British government, inside the walls of the main hospital."

Although he was a colonial physician second class, Yersin did not treat directly the sick (and would have been forbidden to do so anyway by the hostile English doctors). This was the first displacement. The second displacement was the fact that although he was inside the hospital, he was *in* his laboratory. The third displacement was that he brought *with him* his laboratory that he had built after many difficulties. Here we recognize the Pasteurian. In the midst of the worst horrors, it was the laboratory that was given first priority.

The article then takes up the results of a different science, not epidemiology but clinical medicine. Yersin is no longer discussing centers of infection or global geography, but symptoms. He is discussing an abstract patient, whose semiological table he sums up. There can be no doubt that what he is confronted with is "bubonic plague": "Here are the symptoms of the disease: sudden outbreak after four and a half to six day's incubation; torpor, prostration. The patient suddenly has a high fever, often accompanied by delirium."

At the end of the clinical picture, he sums up in a word another set of data, which this time come from hospital statistics: "Mortality is very high: about 95% in the hospitals!" At this stage Yersin has already set up his laboratory and covered three disciplines. From now he completely changes register and moves into what we might call urban hygiene. The city considered as a sick body is studied as a whole: "It is interesting to observe that, in most of the city where the epidemic broke out in the first place and caused most devastation, a new system of drains had just been installed." The circulation of microbes and of dirty water or excrement is as contradictory as that of empires and epidemics. How can one design drains so that they evacuate refuse without contaminating, when those flows of water, excrement, and microbes do not move in the same way? This is a problem that occupied experts' minds for a good half-century.[31] Yersin had no trouble demonstrating the bad design of the drains that evacuated some water but allowed the microbes to proliferate.

He then passed to a new set of concerns, which might be called ethnographical. He meticulously describes the lavatories and the emptying of the soil tubs, "whose contents, after undergoing some preparation, is used to fertilize the innumerable Chinese gardens that border the Canton river." All these details matter when we are considering a city as a culture medium likely to encourage or attenuate the action of a microbe.

But so far Yersin is summing up what he has learned from others. He adds nothing to it. He does not cure anyone, he does not alter the hygiene of the city, he does not rearrange the drainage system, he does not add any new symptom to the clinical picture. Distrusted by the English authorities and competing with the Japanese bacteriologists, he crosses through all these interests. It is the same when he passes through what might be called the social question: "The lodgings occupied by Chinese of the poorer classes are everywhere filthy hovels, which one hardly dares to enter and in which an incredible number of individuals are crammed . . . One can understand the ravages wrought by an epidemic when it takes root in such a terrain, and the difficulty there must be in eliminating it!"

The social question is seen by Yersin in this article only as a terrain for the epidemic. He is not there to weep over the "poor classes," any more than he is there to treat the sick. This does not prove that he was "heartless," only that the program on which he was working required that he spend as little time on the social question as on clinical medicine or hygiene. This Pasteurian program required him to cross through all these disciplines as fast as possible.[32]

So far, Yersin seems to superimpose disparate elements. But the link begins to emerge, a link that only a Pasteurian could consider making. He observes with interest that the European houses are hardly affected. More interestingly still, he writes: "These European houses are nevertheless not exempt from all danger, for one often finds in them dead rats, indubitable clues to the close proximity of infectious germs." We may compare centers of endemic diseases with centers of epidemics; we may compare the various districts in the city according to race, housing construction, wealth, agricultural practices, and drainage systems. But above all—and this is an addition that a hygienist would not make—we may begin to compare the various animals that have fallen sick from the plague. They do not all die, but they are all affected in a different way. Here again Yersin, like Pasteur, learns from the knowledge of his predecessors: "The physicians of the Chinese customs who had had the opportunity of observing the epidemics at Pei-hai and Lien-Chu in the province of Canton and M. Rocher, French consul at Meng-tsu, had already noticed that the scourge, before striking men, began by killing off large numbers of mice, rats, and buffalo." This curiosity of the observers is retranslated by Yersin into "variation of virulence." The districts, races, cities, and animals are each infested by the virulence in a distinct way. That which would be an indecipherable puzzle for any other profession is precisely what enables Yersin to say with some satisfaction: "The particular susceptibility of certain animals to contract the plague allowed me to undertake in good conditions an experimental study of the disease." This sentence is not marked by any cynicism. The plague presents itself well to the eye of the researcher: that is all. The successive preoccupations that he has aligned according to the Pasteurian tactics are all translated into the single language of the variation of virulence of a single organism in terms of terrain.

Yersin now has to define the organism that will replace the social question as it has replaced clinical medicine or epidemiology: "Everything suggested that we had first to discover whether a microbe existed in the patient's blood and in the bubonic tumors." Back in his laboratory, Yersin is interested in the patients who lie there, but in order to capture the microbe he returns immediately to his laboratory. The tumor is no longer a symptom of clinical medicine. It is what must contain the microorganism:"The pulp of the tumors is, in every case, filled with a veritable purée of a short, thickset bacillus, with rounded ends, which is fairly easy to color using aniline dyes, but is not stained by Gram's method. The ends of this bacillus color more strongly than the center so that it often presents a light space in the middle."

It is no longer between Lan-Chow and Pei-hai that a center of plague infection is defined; it is no longer between the fever and the tumor that the symptoms of the plague are defined; it is between aniline, Gram's method, and the microscope. New trials produce a new agent.

But again Yersin does not stop at this actor. He does not write to the minister to describe the bacillus, soon to be known as "Yersin's bacillus." He does things to it in the culture medium that must imitate what the bubonic virus does on the body of the patient: "The pulp of the tumor, when sown on gelose, develops white, transparent colonies, with irridescent edges when examined under reflected light."

In presenting a bacilli culture, a recent advertising headline ran: "The new French colonies." It was intended as a joke; it was also correct. The Minister of the Colonies was to take an extraordinary interest in Yersin's colonies. Isolated in his laboratory, he worked on microscopic colonies in an effort to transform those of the macro-parasite whose "possessions" were threatened.

Following a Pasteurian program, Yersin immediately modified the culture medium, for he was no more interested in the culture than in the bacillus itself: "The culture does even better on glycerized gelose. The bacillus also grows on coagulated serum. In the medium the bacillus has a very characteristic appearance, very reminiscent of the erysipelas culture: a clear liquid, with curds along its walls and at the end of the tube."

It is all there. We can see the centers on the map of China; we can see the poor classes in their hovels; we can see the tumors on the armpits of the sick; we can see the dead rats in the houses of the whites; but even better, we can see the curds along the wall of the tube. Certainty grows when the judgments of perception become simplified. The five photographic plates that accompany the article show neither the Chinese nor the sores nor the dead nor the rats but the colonies under a microscope.

Nevertheless, to vary the bacillus in cultures was of no more interest to Yersin than was any other stage. What he wanted was to give the plague back to the animals. He wanted to reenact an epidemic in his laboratory, which would imitate the one raging outside the laboratory: "If one inoculates mice, rats, or guinea pigs with the bubonic plague pulp, one will surely kill those animals ... The guinea pigs died within two to five days on average; the mice between one and three days." Inoculation imitates contagion. And Yersin now invents a clinical medicine that is no longer looking for symptoms in the patients of the hospital around him, but in the guinea pigs that he deliberately makes sick. He is doing clinical medicine, but on his own terrain: "[At the autopsy] the intestine and the surrenal capsules are often congested, the kidneys purplish, the liver enlarged and red; the very fat female rat often shows a sort of eruption of tiny miliary tubercles."

He does not stop at an experimental anatomopathology any more than at any other of the techniques that he has brought together in the course of the article. As soon as he is in possession of a laboratory epidemic, he varies its virulence, not by modifying the district, sewage system, or race, but by mixing the culture media, the type of bacillus, and the animals: "One can easily pass the disease from guinea pig to guinea pig with the help of the rat pulp or blood. Death arrives more quickly after a few passages." Conversely: "Pigeons do not die when inoculated with a moderate dose of the tumor pulp, or with a culture of the plague bacillus."

In his temporary laboratory, now dominating the epidemic that dominates outside, Yersin concludes: "The plague is therefore a contagious and inoculable disease. It is likely that rats are its principal vehicle."

But scarcely has he concluded than he rushes to the culmination of what appears to him to be the sole justification for all his successive interests: the vaccine. The article does not get so far. Yersin seizes only upon the patients spontaneously cured, and he immediately places the weakened agent that is its cause under the microscope: "By sowing the pulp of a ganglion taken from a patient who has been convalescent for three weeks, I was able to obtain a few colonies absolutely devoid of any virulence, even for mice."

And he concludes: "These very suggestive facts allow me to suppose that the inoculation of certain races or varieties of the specific bacillus would no doubt be capable of giving animals immunity against the plague. I have begun this line of experiment the results of which I shall publish later."

We do not have to read thousands of pages written by hygienists and physicians to appreciate the trenchancy of such a style. Whenever

Yersin speaks of a group of agents, he immediately orients it along a project that does not interest the other agents. A hygienist might read the beginning of this article, but what would he make of the pigeons, aniline dyes, and "purplish" kidneys? A physician might be interested in the last item, providing that the kidneys belonged to a man and not to a guinea pig. But what would he make of the over-narrow drains of Hong Kong, bacilli in gelose, or the towns of the high Chinese plateaus? A social reformer might be interested in the hovels of Hong Kong, but this does not mean that he would enter a laboratory. A biologist might be interested in this new bacillus, which has a light spot in its middle, but he would not know what to make of all the other agents, which would seem to him to be quite incongruous.

Yersin himself was interested in all the series of agents: the macroscopic agents (geographies, Chinese cities) and the microscopic agents (bacilli and their colorings). He followed the human agents just as much as the nonhuman ones, brought his attention to bear on rats as well as on Chinese hovels or European houses. He was as interested in the city as in ulcers. Was he interested in everything, then? No, he was interested in nothing, or almost nothing, for in each agent he took only what might link it along an obligatory passsage that led him, by forced stages, to vaccination. This double movement—using all the agents, making use only of those that led to vaccination—explains his spare, nervous, rapid style, which rests on everything but never stays still. Playing on all the professions, he is always ahead of them, moving each of them by the combined force of the others. In his laboratory, bending over the colonies that he has obtained from the ulcers of patients in Hong Kong, Yersin was to offer the Ministry of the Colonies the plague bacillus, just as Pasteur had given it the agents of anthrax or rabies. Yersin was not involved in politics, he did not treat patients, he did not help the poor, he did not rebuild the drains, and he did not advise the Europeans, yet he moved the positions of fleas, rats, colonial administrators, army doctors, Tonkinese, the poorer classes, and bacilli.

The *Annales de l'Institut Pasteur*

When I take not an isolated article but the 1,500 or so articles that the *Annales* threw into the battle from 1887 to 1919, I find once again Pasteur's use of movement and strategy—and Yersin's style. If we are right to consider a scientific article as a machine, how are we to describe a periodical, the official periodical of the Pasteurians? Along with the lectures that taught investigators from all over the world the skills that were indispensable to the very existence of the phenomena, with the vaccines, serums, incubators, filters, diagnostic kits, and analysis sheets that enabled the laboratory to extend its power, the journal was the most important of the Institut's "products."

Reading this scientific journal, we nevertheless do not leave the solid terrain of trials of strength. On the contrary, we arrive at it. The so-called technical article does not float over laboratory experiments like some empyrean. It is part of the action (Callon et al.: 1986). It is the action itself, the action that constructs credibility and makes the "scientific facts" disputable or indisputable. The only difference from the articles of the *Revue Scientifique* is that the *Annales* are addressed to other laboratories engaged in the same techniques. We may expect, therefore, to see an increase in words, terms, abbreviations, that refer to the local folklore and to the tacit practices of a professional group increasing in size and coherence.

In fact, the overall corpus of the *Annales* is astonishingly faithful to the Pasteurian spring and remains easily analyzable (see fig. 2, p. 268). This is true, to begin with, of the very existence of the Institut. Pasteur began to treat Joseph Meister on July 6, 1885; on March 18, 1886, at the Académie he posed the question of a vaccinal establishment; three months later the Crédit Foncier held an account amounting to 2,586,000 gold francs subscribed by those who trusted Pasteur; on March 14, 1888, the laboratories of the Institut were opened, the *Annales* having begun a year before. Has credibility often been converted into capital so quickly in the history of the sciences? Despite the low number of patients affected by rabies, despite the polemics on treatment, the trust of the public was converted, by the short cut of gold, into laboratories of fundamental research with which it would be possible to produce new facts, reverse other balances of forces, move other social groups, create other agents, extend other networks. In this acceleration of conversion we recognize the type of displacement so typical of Pasteur.

This capitalization, it has to be said again, was not necessary. Pasteur could have cashed in on the public's trust; he could have simplified the research program according to the demands of medicine alone or, on the contrary, have devoted it solely to biology. Instead, we see in the *Annales* a continuation on the same wide front of disciplines and skills that he had initiated but which could not be confined to any definition in terms of profession. In those articles there is talk of cheese, beer, and wine, but also of enzymes and nitrogen, and of the sources of the Seine, which contained bacteria, and of phagocytes and precipitins, and of the wounds of tubercular or diphtheria patients in the Children's Hospital, and of the mosquitos on the Pontine Marshes or of rat fleas in Madagascar. Like Yersin's article, almost none of the articles in *Annales* stops short at the homogeneous set of agents. When it does do so, the juxtaposition of the articles by number or year reproduces these wide-ranging interests.

The journal is neither medical nor hygienist nor even biological. One number of the *Annales* mixes concerns that all the other professions separate and inserts everywhere results acquired in the laboratory to which the other professions are either indifferent or confined.

The same issue, for instance in 1895, comprises articles on the disinfection of feces, on diphtheria, on poison, on the entry of intestinal microbes into the general circulation, on the dosage of alcohol, on the modes of resistance of the lower vertebrates to artificial microbic invasions, on the migration of calcium phosphate in plants, and on the practices of microbic colorings, but also a homage to Pasteur, statistics from the Municipal Antirabies Institute of Turin, and an article on contagion through books.

A journal of hygiene, even after 1880, would be confined to urbanism, or sanitation, or the poor laws. A journal of entomology would confine itself to describing the life cycles of mosquitos. A journal of immunology would speak only of the body and its reactions, without concerning itself with microbes. A medical journal would carefully describe the symptoms or remedies of a disease. An administrative journal would attempt to set out clear regulations concerning the removal of refuse or the burying of corpses. The *Annales* speaks of all these things, passing through every profession, and each time adding enough laboratory results to allow all the professions to continue in their tasks. Weaker than each profession, the Pasteurian was to become stronger than any of them. It was neither lack of intellect nor lack of courage nor lack of laboratory equipment that limited the field of activities of the various disciplines traversed by the Pasteurians; it was only the agent or type of agent that they privileged and the form of network along which they made that agent run.

In order to understand the "scientific" content of the *Annales,* we must understand the originality, which I do not hesitate to call "political" or "sociological," of its authors. The Pasteurians were no more numerous, no more brilliant, no more rigorous, and no more courageous than the others, but they followed a different agent, the cultivated-microbe-whose-virulence-they-varied. With such an agent, they (and it) could ignore the categories on which nineteenth-century society had been built. The hygienists were interested preferably in *external* agents on a macroscopic scale: cities, climate, soil, air, and all *social* agents, such as poverty, overcrowding, and the laws governing commerce. The doctors were interested in *internal* and above all *individual* practices, such as constitutions, terrains, humors, and wounds. Between the external and the internal, the crowd and the individual, there was little contact. The biologist or physiologist was concerned with internal agents that were sometimes microscopic, sometimes functional, which had no necessary relation with physicians, still less with hygienists. They spoke of organs, of the glycogenic function of the liver, of breathing, of cells. This was a huge world that had nothing directly to do with the doctor-patient relationship or the sanitization of cities. Following his attenuated microbes, the Pasteurian could pass from the preoccupations of one of these three great groups to those of others. Of course, the macroscopic or external agents did not interest him as much as they interested the hygienist, but he could use them to understand the movement of a microbe in the human body. The internal individual agents of pathology were of less concern to him than they were to a physician, but he could use symptoms to understand the patient's body as a particular culture medium. The internal machinery of the body was of less interest to him than to a physiologist, but he was able to use it to understand the dazzling progress of the microbe in its economy.

By manipulating his agent, the Pasteurian was always slightly displaced compared to his colleagues, whose discipline he retranslated and sometimes redirected. Always outside, like Yersin, at the moment when the physician was inside with his patient, there the Pasteurian was suddenly inside, bent over his microscope, while the hygienist was confronting the problems of bad housing; and he was suddenly at the hospital bed treating diphtheric children while the biologist was isolated in his laboratory counting or classifying bacilli. We saw how society had been made without taking microbes and the revealers of microbes into account. But in these last years of the century the framework of society was redefined in order to *make room* for the microbes. Elimination of them from the social relations that they distorted also made room for the Pasteurians. The revealer of microbes could easily reshuffle the cards and increase the power of what in reality was a very modest workforce.

This increase is all the more effective in that the *Annales*, like Yersin, refrained from speaking of all the agents that it had linked together along its continuous networks. It was concerned not with poverty in general but with that obligatory point of passage which made the sputum of the poor tubercular patient the direct cause of the contamination of those around him. It was concerned not with administration in general but with those regulations that, by ordering the drying up of stagnant waters, prevented mosquitos from laying their eggs. It was concerned not with pathology in general but with false membranes taken from children's throats that, once cultivated, would make a definite diagnosis possible. It was concerned not with anthropology in general but with the cultural habit that allowed plague-bearing fleas to pass from the rats that they parasited to the blankets on people's beds. The articles in the *Annales* are to be recognized above all, then, by their independence from the divisions practiced in society. They make no definitive differences between hygiene, society, medicine, biology, and industry, nor between chemistry, zoology, and microbiology. But instead of speaking about everything in a vast synthesis, they speak *only* about the agent that they can retranslate into the language of their attenuated microbe. They are thus able, without dispersing themselves, to bring all their efforts to the few points of hygiene, biology, administration, and pathology on which the laboratory allows them to be strongest. By using the microbe whose virulence may be varied, they are able to pass from one discipline to another and move in a single movement from contagions to phagocytes, from these to cheeses, and on to diastases and drains. These sudden changes of scale enabled them to carry off this double coup: they were able to renew medicine *without ever taking disease as an object of study* and to renew politics and hygiene *without ever taking the poor or social outcast as a unit of analysis*.[33]

Indeed the contents of the *Annales* include a large number of articles about diseases, but they are all written in the particular style that is defined here. Only ten of them might be regarded as typical of medicine with its homogeneous agents. Those few articles were all written by doctors invited to describe for the *Annales* the symptoms of a disease that was needed by some other Pasteurian to advance his program.

The Gradual Drifts of the *Annales*

Despite this fidelity on the part of the *Annales* to the spring of the Pasteurian drama, articles like those of Yersin, capable of dealing in six pages with a disease as complete and famous as the plague, from epidemiology to vaccine, are not common. As I have shown, Pasteur links together contingent elements according to a principle of movement that is justified only by success. His successes and those of his disciples should not allow us to forget that, although Fortune smiled on the audacious, she seldom gathered together a favorable set of circumstances. There was of course the case of rabies, but there was also tuberculosis, a vaccine for which did not arrive in the suffering world until 1927. If we read the *Annales* attentively, we see how rare Pasteur's blitzkrieg was. Indeed, to pasteurize a disease was no easy matter. It had, so to speak, to be laid along a curve each stage of which had to be carried out in turn: a link had to be made between a disease and a microbe (and sometimes the clinical picture had to be shaken up somewhat); after that the microbe had to be isolated, a process that was not always possible (if, for instance, it was a virus); then the microbe had to be cultivated in a favorable medium in such a way as to increase its effects, an operation that was often impossible if the wild microbe refused to allow itself to be domesticated; the next stage was to find a laboratory animal able to contract the disease, another difficulty that could seldom be overcome; it was then necessary to multiply the movements from animal to animal and from culture to culture in order to attenuate or increase the violence of the microbe and thus produce a vaccine or a serum; after that the vaccine or serum had to be able to be produced in large quantities and in a stable form; lastly, the distribution of these products had to be extended by setting up or supervising institutions, legislation, industry, and the authorities necessary to this extension. In addition to all these tasks, it was no bad thing to apply them to important diseases. It was also essential that the paternity of the discoveries should be attributed to the researchers of the Institut in order to strengthen the secondary mechanism.

Needless to say, the articles capable of carrying out such a journey in a single go may be counted on the fingers of two hands. Such articles deal with little-known animal diseases. What we find over and over again in the *Annales* are examples of a partial Pasteurian program. There may be a microbe, but no disease to go with it; a microbe and a disease, but no medium to cultivate it; a culture, but no animal to take the disease; diseased experimental animals, but no way of making a serum vaccine. It is as if the diseases themselves had their own history and did not allow themselves to be trapped by the strategy of the Pasteurians as easily as the anthrax bacillus or plague bacillus. So innumerable articles are required to speak of the partial stages of the program. Furthermore, detours become longer and longer. One article speaks not of the culture of the plague bacillus but of the coloring method; another speaks not of animals sick with the plague but of experimental diseases with laboratory microbes that exist only there.

To this differential resistance of the diseases and this extension of the detours that sometimes justified an entire career, we must now add the increase in the numbers of researchers themselves. At the turn of the century there was now a whole crowd of them, abroad as well as in France, working on the infectious diseases. This crowd brought with it a division of labor, and it therefore became difficult to recognize the strategy of a single article. To follow one disease, it was now necessary to read dozens of articles. This subdivision and fragmentation had one great result: the researchers were now able, like the microbes, to be multiplied in the laboratory and make their careers without needing to plan their efforts explicitly according to the Pasteurian spring. In other words, the presence of external factors within the laboratories was to become much less clear. A pocket of techniques, skills, and private language could now regard itself as relatively isolated from the other disciplines. "Science" was to appear in greater isolation, and correlatively, the analyst was to have more difficulty explaining it. In speaking of isolation, I am not questioning the principles from which I set out. Isolation is just as material and just as real a strategy as the Pasteurian movement and recruitment. It is simply either less skillful, because the interest of the other agents is lower, or more skillful, because it no longer needs the other agents.

We may follow the creation of this pocket and the movement of interests both in the *Revue Scientifique* and in the *Annales* (see fig. 3, p. 269). After diphtheria (1894) and above all after the passing of the great law on hygiene (1902), the Pasteurian revolution no longer hit the headlines. The reader was referred to the innumerable scientific journals that had been founded to speak about it. The headings multiplied: "parasitology," "immunology," "microbiology," "tropical medicine," "biochemistry," "serotherapy," "social hygiene." The great authors no longer wrote directly in the *Revue* if they wanted to speak about hygiene. The short circuit of the Pasteurian spring no longer worked, at least until the Pasteurian reappeared as the Solon of the Tropics. A huge pocket that first crushed the study of infectious diseases and then almost completely disappeared when the parasitical diseases were reached. How is this pocket, which seems to occupy so much room, to be explained?

If the reader had admitted that first Pasteur, then his disciples, linked together interests that without them would have gone in different directions, he would not be surprised that the slightest uncertainty as to the direction of the movement should immediately divert the Pasteurians themselves. Champions of movement and translation, they too could, if they softened, allow themselves to be interpreted and used. This is what happened with Metchnikov's work in the Institut. It is the story of the translators translated and therefore betrayed.

The microbe, the agent in the drama of infectious diseases, was obviously "evil." But what would happen if, instead of looking at the action of the evil entity, they looked at the body's reactions? The laboratory would become populated with a crowd of new agents, the most famous of which were the phagocytes, the specific enemies of the "evil agents." But in a second movement these phagocytes did not always react to the complete microbe. Part of the microbe, an isolated toxin, might bring on both the disease and the reaction. But what the microbe or toxin gave rise to could be brought on by *anything:* dust, an eyelash, a chemical product. This third movement deprived the microbe—as it had been defined by Pasteur—of any important role. Immunology could stick to bodies between flesh and skin, give itself antigens at random, and provide work, solely within the walls of the laboratory, for dozens of researchers for years on end. There was no longer any need for microbes, nor for stories of wolves and foxes with rabies, nor for poor Chinese, nor for disinfectant control. The microbes were becoming particular cases of a general problem: the integrity of the organism.

This movement from the initial program is all the more important in that elsewhere at the Institut the microbe and its variations in virulence were no longer taken as the unit of analysis. Duclaux's program took the microbe from below. The "microbe" is not a definitive, obvious, natural agent. It existed for a time, *in the absence of anything better,* before it in turn was distorted. These successive distortions show clearly how wrong it is to speak of "discovered microbes." Making the microbe act as a unity is, for Duclaux, a temporary arrangement of conditions.[34] It is much more interesting for him to break the microbe into its components and to make each of them work in turn. The biochemistry program took on an enormous importance in the *Annales* and completed the destruction of Pasteur's microbe that immunology had begun.[35] The *Annales* deconstruct the microbes whose-degree-of-virulence-may-be-varied-at-will and construct with its dismembered elements immunology, on the one hand, and biochemistry, on the other.

To these two great movements, which deprived the attenuated microbe of the unity of action that it had enjoyed for only a few years, we must add the accumulation in the laboratories of strains, colonies, archives, which made it possible to produce new agents without moving outside the laboratory. This movement took place so to speak on the spot and was the result of the very accumulation of artifacts.

If we add up all these succesive constraints, without of course forgetting the stubborn resistance of the diseases themselves, we can explain many of the overall movements of the *Annales*. The complete Pasteurian program was respected at the beginning, but always partially and over a limited number of diseases. The classical programs, such as hygiene or micrography, which corresponded to the disciplines begun by Pasteur, declined and disappeared. The programs begun by Duclaux and Metchnikov grew, but even they displayed only fragments of the initial direction. Only Bordet's immunology was situated once again in the line of classical Pasteurism. In Bordet's hands, immunology became a rather long detour toward new methods of diagnosis and treatment. But it was Laveran's program, which actually concerned diseases without microbes, that finally reenacted the typical movement of the Pasteurians, by which the entire planet was ultimately to become the field of action of the Pasteur Institutes.

If we made a cross-section of the *Annales* in 1894 and again in 1910, we would find all the elements of the Pasteurian impetus. But if we took a cross-section in 1900, we would have more difficulty finding that proliferation of heterogeneous agents recruited by the Pasteurian strategists. We would be more aware that we were reading "scientific articles independent of social concerns." Metchnikov had other allies, other aims, than Pasteur; he worked only in the laboratory, constructed agents between flesh and skin, and indeed actively isolated himself and his students. The word "isolate" does not have any ontological sense. It does not trace a sacred boundary. If we speak of isolation, we must give that word as material a meaning as the fabrication of an isolate in fiber glass or double glazing.

We can see how powerless our analysis would have become if we had left to one side the "technical" articles by Pasteur and others in the *Annales*. We would have understood nothing about the society of the time, nor anything about the content itself. Even the impression of a separate content is due only to such techniques of isolation as those of Metchnikov and above all to the attribution of responsibility, which limits the number of allies acquired by a science to just a few.

We understand also why the Pasteurians were always so virulent in attributing responsibility. With so many allies, they ran a great risk. They ran the risk of confusing their allies in a mass of anonymous workers, many of whom did not even wear white coats. To create "science" and "Pasteurian science" out of this crowd of agents, nothing in the secondary mechanism was to be ignored. Speaking of one of Pasteur's predecessors, Duclaux writes: "Davaine did not even understand his own discoveries" (1879, p. 629). What he did, therefore, he did not really do. However, there were predecessors who seemed indisputable because they had been born a century before, such as Jenner, inventor of the vaccine. Even these, however, might be resituated in time. Without Pasteur, the inventor of a general law of which Jenner's vaccine is merely a particular case, Jenner would not even be a precursor, says a disciple. From the point of view of the secondary mechanism, Pasteur is *before* Jenner, for he provides the foundation for him as much as for the German Koch, whose role is enormous in all the non-French histories of bacteriology. It was Pasteur himself who pleaded for the attribution of responsibility. How harshly he reminds Koch of the order of primogeniture—"you had not been born to science when I was cultivating microbes"—and to end the discussion, he makes use of the result of another argument over priority won against Lister, with Lister's consent: "These skillful and courageous experimenters place the starting point of their studies in a reading of my memorandum on putrefaction" (1883, p. 74).

But times were hard. We keep a chronology only as long as we can act on people. When an American wrote a history of recent developments in biology that paid no attention to the secondary mechanism set up by Pasteur, the *Revue Scientifique* published the article but demolished it, restoring the chronology and true responsibilities. Richet settles the matter with this summum of absolute idealism: "M. Pasteur has really been the soul of all this science, *principium et fons,* and it is really not enough to say that before Koch no name could be set against his. It must be said and repeated that before and after M. Koch, no one can be compared with Pasteur" (1889, p. 330). By a series of similar battles against the Germans, Americans, and British surgeons, but also the Lyonnais, biologists, physicians, technicians, and other dear colleagues, it was possible in the end to achieve an isolated Pasteurian science that became the cause of social transformations. This mysterious efficacy attributed to the "sciences" and to Pasteur's genius is, like other mysteries, an interesting construct, whose analysis should present no difficulty, at least in principle.

Chapter 3
Medicine at Last

Why Tolstoy

Times Are Hard

So far I have shown the immense forces that made up Pasteurism, and the evolution of Pasteurism toward the end of capturing these forces, supporting them, translating them, and attributing their power to itself. My demonstration has had the inconvenience of considering only successes. Hygienists and Pasteurians were moving roughly in the same direction. Tolstoy, my inimitable model, had the skill to choose a range of winners and losers, with a view to testing his various hypotheses concerning the make-up of the forces. Similarly, in order really to judge the fruitfulness of my analysis, we would have to find control groups whose behavior, during the same period and on the same questions, was totally different. Only then would we see clearly that the "evidence of reason," the "force of logic," the "thrust of progress," the "power of interest," the "irresistible technical efficacy," the "ripeness of time," are merely victor's words. They are cries uttered in the midst of battle in order to ward off fate, as Tolstoy shows with the cries of "rout" or "victory" at Borodino.

The first way of multiplying control groups would be to tell the story, not by groups of social "actors" but by groups of "nonhuman"

111

agents, and to look a little at the diseases themselves. Few diseases obey the fine ordering of irresistible progress that renders them definitively a thing of the "past." The symbol of this "resistance" on the part of diseases, whose rhythm does not obey that of the groups who announce their disappearance, is of course the Spanish flu (Katz, 1974). World War One was, in the opinion of all, a triumph of modern hygiene. Without the bacteriologists, the generals would never have been able to hold on to millions of men for four years in muddy, rat-infested trenches. These men would have died before gas and machine guns had carried them off. This war was the first in which one could kill immobile masses, because hitherto in history microbes had always done the job better. After this triumph of bacteriology, the Spanish flu wiped out some fifty million people in 1919 without the Pasteurians being able even to identify the agent.

The same setbacks occured in any year chosen at random, as shown by the way various diseases were discussed in the *Revue Scientifique*. Let us take, for example, the year 1893, Armaingaud discusses tuberculosis at length but declares that science has ended its work; that is, the disease has been linked with Koch's bacillus, but he makes no attempt to pasteurize it. His aim is to increase the number of Leagues for the Advancement of Sanitoriums (Armaingaud; 1893, pp. 33–42). Now nothing is less Pasteurian than a sanitorium. Here, then, is a disease that Calmette was to take decades to catch up with and whose history is absolutely different from that of anthrax or typhoid fever. Yet in the same year there is a discussion of tetanus in terms of a disease so pasteurized that a serum has been found for it. But when on April 30 of that same year there is a reference to typhus, it is to say that the disease defies analysis. A convinced Pasteurian, perhaps Héricourt, even confesses his leaning toward heresy: "We will have to seek the influence of the direction of the prevailing winds, of drought, and of humidity, etc. We will almost have to go back, and let us have the courage to say so, to the hypothesis of our forebears concerning the influence of the stars" (Anon.: 1893, p. 539).

Here is proof that time does not pass. It has to be made to pass, disease after disease, social group after social group, without which it just moves off in the wrong direction.

When we speak of smallpox, which has been pasteurized for a hundred years, it is to salute the victory in England of the leagues *against* cholera: "The time is past when we could hope that revaccination would become so common in England that it need not be

Absence of Smallpox/Vaccination

made obligatory. Not only has the movement for revaccination been stopped, but vaccination itself is carried out less and less and on a great many points the law that imposes it is no longer obeyed" (Anon.: 1893, p. 699).[1]

The attempts by the hygienists to align laws and science in the same direction and to turn smallpox into one of those great antediluvian species was so fragile that time turned back. We go from the absence of smallpox to the absence of vaccination.[2] Fortunately, 146 people died of smallpox in Leicester where the antivaccination leagues were at their strongest, which strengthened the weakened position of the hygienists in the nick of time.

Tuberculosis and smallpox have their own histories. So, too, does cholera. In that same year, 1893, it was mentioned only to describe, in the purest hygienist tradition, steps taken to observe the pilgrims going to Mecca. Yet in that same year the results of treatments for diphtheria were published. The treatments were a complete success. This disease was soon to be completely pasteurized, after a movement in the program that Pasteur indeed could in no way have anticipated.

According to the disease under consideration in the *Revue*, time stood still or moved forward. Depending on whether there was dispute or not, progress moved forward or backward. There was here a variation in interest and conviction that is as interesting for sociologists and historians as the variations in virulence were for the Pasteurians. Imitating their method, I shall select a series of control groups and see how differently they reacted to Pasteurism: army doctors, ordinary civilian physicians, colonial doctors, and the ordinary people on whom, in the last resort, all the others descend.

The Army Doctors

To be faithful to my initial principles, I should explain *with the same arguments* what stopped the doctors and what made the hygienists rush in. It is pointless to say that the hygienists acted and the doctors resisted, that the hygienists were mature and the others not, or to use another metaphor that serves as a refuge for ignorance, that the hygienists were "open" and the others "obscurantist." Without competing with the work of Jacques Léonard (1967, 1979, 1981, 1986), I would like to show through a study of the *Concours Médical* how the rejection of these asymmetrical explanations enables us to explain the making of time. From the moment we reject the diffusionist model, which attributes the power to revolutionize society as a whole to Pasteur's genius alone, we must, as I have already shown, attribute to the individual who seizes upon a discovery as much and sometimes more power than to the individual who proposes the discovery. That is why I had to speak so much about the hygienists before I could turn to the content of the Pasteurians' programs. But conversely we cannot explain why a discovery does not spread simply by the resistance of various groups. We must understand why they did not seize upon a particular discovery. Those who "accepted" and those who "rejected" it are both agents of that society, and in order to understand their movements, we must neither admire nor blame them but seek to understand the principle of their activity from within.

In order to demonstrate this point, we have one great advantage. We have at our disposal a control group within the control group. Indeed, army doctors seized upon Pasteurism with the same avidity as hygienists. From 1881 Alix realizes the formidable lever that gave his profession the Pasteurian model: "Public opinion is beginning to be moved by progress in the sciences of life." If it was moving, we could take advantage of it and move it forward and thus advance our own affairs a little. Why was Alix so interested? He explains: "It is impossible to deny that, in the very near future, medical questions will all be resolved by applications deriving from the discoveries of hygiene; pharmaceutical therapy in civilian public medicine will fall to a very secondary rank, as has already happened in military medicine" (1881, p. 761).

In giving Pasteurized hygiene a push, our Alix raises the status of his own medicine above that of his civilian colleagues. Furthermore, military medicine had already been pasteurized institutionally. After all, have not barracks always been an ideal laboratory where legions of healthy young men are subjected to a uniform regime? There is no "doctor/patient relationship" with the recruits, who are marched through their medicals and inoculated in batches. This somewhat crude medicine without patients is now seen by Alix as an anticipation of the future of medicine itself. Is it surprising that someone who has everything to gain from such an innovation should seize upon it with a view to extending its effects?

But to understand the activity of these army doctors, it is not enough to believe that they were looking for "legitimacy." This vague word deriving from sociology almost always hides the real content of actions. Their essential problem was that men died in the barracks in peacetime. If we do not jugulate epidemics, someone else writes, the "nation will be afraid of the barracks where it sends its sons and from whence not all return" (Anon.: 1881, pp. 72–78).

There was something more serious still. In wartime, as is well known, there are more deaths from microbes than from the enemy. The conflict between "Napoleon" and "Kutuzov" is duplicated, according to Cartwright, by a conflict between "General Napoleon" and "General Typhus": "The French marched on Moscow without encountering opposition, but typhus marched with them. That army of 100,000 men lost 10,000 soldiers from disease, in the week 7–14 September alone" (1972, p. 97). Every association is transformed by this battle on several fronts: "Fear of the Russians and of the vengeance of the Poles made the soldiers keep together in compact groups. The fleas from the hovels moved everywhere, sticking to the seams of clothes, hair, and brought with them the typhus microorganisms" (p. 92).

In 1802 a French army set out for San Domingo with 58,545 men. In four months 50,270 were dead of yellow fever. In 1809 only 300 of them remained, and they were repatriated to France. Lémure, an army doctor, describes in 1896 the Madagascar expedition: "The Hova government was counting on the fever to prevent our soldiers from getting to Antananarivo. It was depending on that weapon much more than on bullets and shells made in England" (1896, p. 47). Without firing a single shot, the Hovas were content to force the French to bivouac in the plain: "Two months were enough to reduce their numbers to a half and even to a quarter, leaving some battalions existing only in name—all sick and 5,000 dead, that's the balance sheet (out of 24,000 men). Which proves that the expedition was above all a business of sanitation" (p. 50).

It is not I, the author with a twisted mind, who sees trials of strength everywhere. The actors studied have taught me this lesson. The Hovas formed an alliance with miasma to win a war against those armed with rifles and canon. In order to reverse once more this balance of forces, which had already been reversed once, what had to be done? They had to use modern bacteriology. By crushing the microbes of parasites in the laboratory, they eliminated the power of the Hovas' allies and therefore gave the canon and rifles back their superiority, since those who used them would no longer die. When you are an army doctor, you can hardly hesitate.

In war there had always been two enemies, the microscopic and the macroscopic. If the doctor succeeded in jugulating the second, far more lethal one, he gained enormously in importance and became almost the equal of those who fought against generals and canon. In a country like the France of that time, inspired by ideas of revenge, obsessed with its falling birthrate, it soon became unthinkable that whole battalions should be lost to microbes against which Pasteur, the great Frenchman, had produced the remedies. Army medicine was converted to Pasteurism without putting up the slightest resistance. This development is not to be attributed entirely to the Pasteurians, but to those army doctors who seized on the Pasteurians and invested massively in them. The army doctors in turn owed their credit to all those who wanted a strong army and of whom the doctors easily became the spokesmen.

The Doctors Find Pasteur Disputable

The ordinary civilian doctors provide the best counterexample, since they were out of step. As Léonard has shown, the doctors were skeptics. Even more than skeptics, they would be called "grumblers," if that category were accepted in sociology. Those who were directly concerned with diseases and patients saw nothing extraordinary in Pasteurism, or even relevant, at least before 1894. When they at last made up their minds to use Pasteurism, they saw it not as a revolution in their own practices but as a way of *continuing* in strengthened ways *what they had always done*. Finally, when they had fully assimilated the interest of Pasteurism after the passing of the law of 1902 on the organization of public hygiene, the new medicine seemed to owe more to the old than to the Pasteurian strategy, which had in the meantime shifted toward tropical medicine.[3]

What the other protagonists said about the hygienists, surgeons, or army doctors defined in absentia the reasons why private doctors did not budge an inch to make use of Pasteurism. In simpler terms, all the progress of Pasteurism amounted to for them was a dissolution of the medical profession. Others criticized private physicians for their obscurantism, whereas they were being asked to commit suicide. What group would do so willingly? The Pasteurian strategy amounted to attacking disease by a transversal movement which never took the individual sick person as a unity. How could that bring joy to a doctor who knew nothing but the sick individual? What could he make of this vision, which was both too public and too biological, without

Never focusing on the patient

ever focusing on the patient? What could he make of a few great infectious diseases, which amounted after all to a fraction of his daily work and which were of such a scope that they lay quite outside the capacity of the local physician? What could he do with all those pigs, chickens, dogs, horses, bulls, broods, that had so little to do with medicine, with a human face? What could even be done with a cure, spectacular as it certainly was, like that of rabies which concerned a very rare disease and which, furthermore, required that a patient go to Paris to be cured by a product that was absolutely unavailable to ordinary physicians? In short, what could be done with all those doctrines and methods that were the negation of medical work? The answer is clear: nothing, or not much. And since there was nothing physicians could do with those doctrines, they expressed a polite but unenthusiastic interest, tinged even with a certain ironic condescension. This proves nothing about the obscurantism of the physicians; it proves simply that the Pasteurians had not yet learned to take those allies, unlike the other allies, the right way.

The *Concours Médical,* a corporate journal if ever there was one, speaks of Pasteur's work with a distance and prudence that contrasts starkly with the avidity of the hygienists, insisting that Pasteur be absolutely right and extend the implementation of his work at once. The "conclusive" character of Pasteur's experiments can indeed not be attributed to their inherent qualities. The doctors found disputable what for the hygienists was indisputable. Of course the doctors showed "good will"; they supported the subscription to the Institut Pasteur and were proud of him: "We feel deep joy at the idea of fighting the good fight as obscure but willing soldiers" (Anon.: 1888, p. 530). But they were cautious: "Of all that slowly accumulating work, a body of precise knowledge will certainly emerge one day; we can just about catch a glimpse of the way ahead and already a great many facts are piling up. But we should maintain a certain reserve for the time being and not see bacteria everywhere, after previously seeing them nowhere. The aseptic method in surgery has already given great service, not so much by its detailed applications as by the correct ideas of which it was often an exaggeration; it will no doubt serve the physician likewise" (Gosselin: 1879, p. 159).

Were they obscurantists? Did they resist? No, they took great care to separate what was exaggerated from what was useful from their own professional point of view. At the time when Pasteur was attempting his takeover of medicine and the hygienists were claiming to have conquered the state because of the added power offered by the Pasteurians, the physicians waited to see how they would get out of a very difficult situation in which they had everything to lose and preferred to maintain the state of affairs that they had set up with such difficulty: "We believe that, despite the somewhat impassioned attacks of Monsieur Pasteur, clinical medicine is not quite dead" (Reynaud: 1881, p. 102). They defended themselves, which was quite normal. They even took a certain delight in giving Pasteur lessons in scientific method: "M. Pasteur ended his communication (on cholera among chickens) by deducing from those various facts applications to the general history of contagious diseases. We shall not follow the learned chemist in his generalizations: before deducing such conclusions from those facts, which are certainly very interesting, he should repeat and vary the experiments" (Anon.: 1880, p. 177). Like Koch and like Peter, the physicians of the *Concours Médical* were of the opinion that Pasteur really was exaggerating. How can we deny that they were right?

How Were They To Defend the Doctor-Patient Relationship?

The hygienists had a great social movement to translate and a great project of transforming the cities that led them to all the sources of power—just like the Pasteurians sent by Science on the conquest of microbes; just like the surgeons who, by following the antimicrobe, could reach at last organs that had hitherto been forbidden them; and just like the army doctors who could develop more rapidly the strength of armies by adopting Pasteurism. The physicians, however, were mandated by no great social movement. They were translators not of public health but of a multitude of doctor-patient relationships. The conflict between wealth and health, which drove on all the other agents, paralyzed the physicians.[4]

They had other conflicts to engage in. The *Concours Médical* reveals in an almost caricatural way a professional body struggling for its existence, fighting against the world. The medical corps, according to the unionized physicians of the *Concours,* was at its lowest point. It was ill-regarded, ill-paid, overworked, and above all constantly threatened by disloyal competition from everywhere else. As the weeks passed, the journal shows our militants fighting against pharmacists who prescribed drugs; against the sisters of charity who, out of religious zeal, took the bread from the mouths of young physicians;

Conflict between Health & Health *(handwritten)*

against health officials[5] whom the physicians had not yet succeeded, before 1893 at least, in ridding themselves of; against the "pharmaceutical specialities" sold in made-up form by industry; against the health societies which persisted in teaching the public how to bind up wounds" (Gassot: 1900, p. 97); against the bone-breakers, spiritualists, and charlatans who competed with physicians even in educated households. No, the life of a physician was an infernal one, and carving out of French society a space where it was possible to treat people for money required a constant struggle. The conflict between health and wealth became for each physician a matter of how to earn a living while treating people.

The *Concours* also fought against the army doctors who had the affrontery to take on private patients; against patients who refused to pay; against judges who always gave judgment against the physicians in favor of either their debtors or those who accused them of negligence; against the press, which gave physicians either a favorable image, thereby attracting students, or an unfavorable one, thereby robbing them of patients; against the rich and fashionable physicians of Paris who despised their poorer, unknown brethren; and against the other professional bodies, which refused to cooperate. In short, physicians of the *Concours Médical* had nothing but enemies, not to mention the fact that progress in hygiene, by reducing morbidity, was further depriving them of patients (Anon.: 1900, p. 79) and that young colleagues, causing another scandal, were crowding into the faculties and increasing competition.

Again, the physicians' interests were no narrower and their mentalities no less enlightened than were those of the hygienists, surgeons, army doctors, or Pasteurians. They were fighting to save a profession and to resist upheavals that were outside their control. Indeed, they were caught up in a paradox that it was difficult for them to escape. The laissez faire of unbridled liberalism would, in their view, allow their enemies to put an end to them fairly quickly. For instance, mutual aid societies could easily guarantee a young physician a fixed salary in exchange for a monopoly. In order to prevent such a takeover by external forces, physicians had to oppose medical liberalism by forcing the young physician to join his colleagues in a union, so as to maintain competition between them. For example, unions forced the mutual aid societies to recognize the patient's "right" to choose "freely" his own physician. For practitioners, the choice was the following: either they did not join the union and "kept their freedom," whereupon medicine as a corporate body would disappear, overcrowding of the profession would spread, and government-employed physicians would have a monopoly over each category of patients—school, tubercular, hospitalized, vaccinated; or they joined the union and thereby effectively prevented the "free"—that is, nonunionized—physicians from limiting free, loyal competition. A free medicine required a determined corporation. Again, this situation is in striking contrast with that of the hygienists, whose power gained from concentration and from being merged with that of the public authorities. The physicians had everything to lose from such a merger.

We can easily understand that with such problems the physician could have nothing more than a polite but distant interest in the acrobatics of microbes in laboratories. Either the physicians could use what was taking place in the Institut Pasteur to advance their own interests, or they could not. If they could, any argument, *however revolutionary it might be,* would be understood, seized upon, transported, and used as soon as possible, as in the case of the army doctors. But if they could not, no argument, *however useful and important it might be* in the eyes of others, could be understood or applied even after a century. The time of innovation is not like a general grid on which one could point out the "resistances" or "maturity" of social groups from year to year. The time of innovation is the *ultimate consequence* of the interests of social groups in one another and in their movement. Innovation takes time if those interests do not coincide or cannot be translated by a shared misunderstanding. It moves very quickly when the forces are pulling in the same direction, as in the case of hygiene, and slowly or not at all when the forces oppose one another. The physicians provided a perfect illustration of this essential negotiation of time. As far as science was concerned, they remained as they were—that is, time was suspended for them—until the displacement of the Pasteurian programs finally aligned an innovation with the interests of the physicians struggling for their survival, as in the earlier case of the hygienists.

The source of the *Concours Médical* throws admirable light on this reversal. But we must go back to the *Revue Scientifique* if we are to grasp how the *other* professions and social movements of the time saw the future role of the physicians. Although an actor is always active, as the name indicates, some actors are defined by others as being passive. This was the case with the physicians until 1894. All the groups expressing themselves in the *Revue* defined the physicians as a passive group requiring radical reshaping, and they laid out in detail what should be done with that profession.

One Agent Turns the Other into a Patient

Of course, everybody showed verbal respect for the physicians. Pasteur always said, "If I had the honor to be a physician," I would do this or that. Yet throughout the pages of the *Revue Scientifique* there is nothing but contempt for that skill "belonging to another age" (which was made to belong to another age by being superseded). This contempt derives from the fact that the physician was regarded as a child fighting in the dark against tiny beings that were secretly manipulating him. Whoever is manipulated by a microbe unknown to himself may in turn be manipulated without too many scruples with a view to putting him on the right path. The *Revue* almost never spoke of the physicians as an active group—indeed the physicians themselves, unlike the surgeons, never spoke of "us"—but always as a passive group. Dozens of articles set out to show medicine the way that it *must* follow, but *none* of the operations proposed for it was within the grasp of the small private physician. Practitioners were shown the way that their art must follow in order to be transformed, but that art was a science known only to laboratory scientists. One anonymous commentator writes naively: "In the past, since we did not know the cause of diseases, there were only patients and the interests of patients involved. Now that we know the external causes of the diseases that are to be pursued and destroyed in the environment as a whole, cosmic and social, the authority and influence of the physician have naturally benefited from such an enlargement of his field of action (1889, p. 630). This "naturally" is valid only for the hygienist, since the physician could extend his field of action only by completely denying what he had done hitherto. The hygienist with the hybrid notion of contagion environment could go on doing what he had been doing while becoming pasteurized. Even the surgeon could carry on with surgery while accepting the premises and first fruits of pasteurism. But if the physician were to become pasteurized, he would have to abandon his patient. If so, what would he do? It would then be pointless talking to a physician of his "well-understood" interest or his "long-term" interests. No agent can change. They can only shift slightly.

The *Revue Scientifique* never tired of declaring the end of curative medicine. For a physician, this was not a very pleasant thing to hear. The entire Pasteurian takeover of medicine was aimed at redefining pathology so that disease would be *prevented* instead of *cured*. Richet writes in a polemic: "Pasteur alone . . . has made more progress in medicine than have 10,000 practitioners more competent than he in medical science" (Jousset de Bellesme: 1882, p. 509). The reason for this progress was simple enough and enthused all the authors of the *Revue* who were tired of medicine. Pasteur's hygiene "makes it possible to prevent the morbid causes, to remove diseases, so as not to have to cure them" (Alix: 1882, p. 149). This claim, which was gradually to disappear before the end of the century, cut the ground from under the physicians' feet. "It is easier to prevent a hundred people from falling ill than to cure one who has become so," writes Rochard (1887: p. 388). How was a social group, the physicians, to be made cooperative while they were at the same time being warned that they would soon have no more patients to treat?

Not only were the doctors despised, not only was disease to be, redefined by removing the patient, and not only was the art to which the physician had devoted his life apparently doomed to imminent extinction, but "they" even wanted him to play a role absolutely contrary to everything he had learned and contrary to his age-old interests. "They" wanted him to declare diseases contagious. I know in sociology of few such good cases of the redefinition by one social group of the role of another group. Role of social group

Hitherto the physician was the confidant of his patients and held to medical secrecy, which was upheld by all the rules of law, propriety, and medical ethics. Now the other protagonists were about to turn these rules upside down and demand that the physician denounce the contagious patients. Nothing could better show what is meant by being *acted upon* by others. The reason for this upheaval is a fundamental one. The Pasteurians added to society a new agent, which compromised the freedom of all other agents by displacing all their interests. The hygienists therefore demanded that microbes be prevented from propagating by interrupting the chain of contagion. The only way to achieve this was to isolate the patient before he could contaminate those around him. But the only way to isolate him quickly was to inform the hygiene services immediately. Only the physician could do this by declaring to the authorities that his patient was ill. But where did this leave medical secrecy? It would be a crime to keep secret the source of a contagion. But what of the physician's role? It was now reversed. He was no longer a confidant of the *patient*, but a delegated *agent* of public health to the patient. But what of individual liberty? The presence of the microbe redefined it: no one had the right to contaminate others. In order to save everyone's liberty, the contagious patient must be notified by the physician, isolated, disinfected, in short put out of harm's way, like a criminal. *Disease was no longer a private misfortune but an offense to public order.* Into the middle of the stage, which had hitherto been occupied by the physician and his patient, there now burst, as in the counting-out rhyme, the microbes, the revealer of the microbe, the hygienist, the mayor, the disinfection services, and the inspectors—all out to track the microbes down. In redefining what made up society, the Pasteurians contributed to this general movement of the authorities—a movement that, like an earthquake, totally subverted the role of one of the agents, the physician.

This reversal, in which the physicians were acted upon by others, is taken as self-evident in the *Revue Scientifique* but gives rise to howls of disapproval in the *Concours Médical.* An anonymous physician warns: "You will then know, if you declare a disease, the squads of disinfectors; your rooms, your furniture, you yourself will be carbolized and your patient isolated; the entire district will be upset, and you and the others will be treated like bearers of the plague; friends will flee you, you will be left alone with your chloral, your carbolic acid, and your patient, who will get no better for them" (1894, p. 212).

Up to 1894 the *Concours* never ceased to inveigh against the dangers of notifying the authorities about diseases, which was regarded by others as an opportunity for the physicians.[6] The physicians worked along networks that were as short as possible; to become an inspector on a long network, whose center was in Paris, seemed to them at first to be the source not of a new power but of a new impotence. We believe only what may be of benefit to us. In the short network that linked the individual patient with the physician, only the *trust* of the patient could be returned. If it was lost, everything was lost.

It was much later that a physician like Valentino could write: "If medical secrecy were abolished, the physician would nevertheless be put in charge of public health·and public hygiene by society . . . he would be allowed to ignore the selfish desiderata of his clientele and become truly what he ought to be: the servant of society" (1904, p. 355).

Between these two quotations a reversal of the physicians' role, took place. From being patients, they became agents. They *became active*. The trust of the patients along a short network became less profitable in the end than the trust of the public authorities, allowing them in return to act upon patients.

In order to understand this fresh start, we must understand that those who redefined the role of the physician needed him. They dictated his new duties to him; they discussed, without consulting him, what studies he should follow; they explained to him in detail what gestures he must make in diagnosing diphtheria; but in thus insisting on the reformed physician, they showed that they needed not a passive servant but a cooperating agent. Although their words were marked by absolute idealism and although they always spoke of "progress" and "diffusion," the hygienists knew very well in practice that they needed to form alliances with active groups if a gesture or technique were to spread into every corner of French society. The physicians may have been despised and regarded as obscurantists or incompetents, but who else could be relied upon to spread hygiene? They could, as in England, have created a new professional group that might have worked, side by side with the physicians, as agents of public health.[7] But in France the authorities decided to use the physicians, the only people at hand, so to speak, with a view to getting them to do what hygiene required of them.

If the physicians were reformed, reeducated, and offered certain satisfactions, they would be quite capable, according to the authors of the *Revue,* of making adequate agents for applying the new scientific and legal rules. To make them "up to it," Rochard claims that they had to be taught the new sciences: "This is not too much to ask of men whose professional training ended at the age of twenty-five." Rochard is the first writer in the *Revue* to fall back on the physicians— in spite of themselves, it seems—as servants of public health, once the illusion of a complete disappearance of diseases began to fade. He defines somewhat defiantly a contract to be drawn up with the reformed physicians: "In the medical army, there are various aptitudes and roles, and when the country places ever more trust in the physicians, it is right to demand in the deliberating assemblies that they extend their knowledge" (1887, p. 390). He adds: "It will be absolutely necessary to give them precise instructions and to make sure that they do not depart from them" (p. 391). The physicians may not have known what they wanted, but there were others who seemed to know for them, and in detail. If they did not understand their own interests, others did not have to be told what they were. The physicians were not trusted, but they were needed. Even in 1894 Richet writes: "The early, definite diagnosis of diphtheria can be established only by the use of bacteriological methods. We must insist that physicians use these methods" (1894, p. 412).

Pity the poor physician—his role redefined by others, robbed of his own definitions of diseases, turned upside down in his medical ethics, made the representative of a new force that at first denied his role and then told him in the minutest detail what he had to do in his consulting room and what methods he must employ. As if such mistrust were not enough, people were being called upon to urge the physicians to conform to the dictates of the Institut Pasteur. I was wrong to say that only the microbes suffered during this period. The physicians did too.

And yet Richet might have refrained from this ultimate sign of distrust. For it was precisely in that year that the physicians seized upon the role that was being imposed upon them, retranslated it, amplified it, and in the end conquered their conquerors.

Where the Patient Becomes Agent

The physicians of the *Concours Médical* were well aware of this redefinition of their role, thrust upon them from above, which was intended to reform them through a contract that from their point of view meant that they would lose everything they had. Speaking sarcastically of the reforms proposed by one prefect, an anonymous physician writes: "This individual is under the impression that, for all the organizations that he proposes to set up, there are collaborators, awaiting orders, scattered throughout a large part of the territory of France" (1887, p. 362). And this was precisely what the others were proposing to do. We thus see from the side of the displaced agents what the *Revue Scientifique* was saying on the side of the displacing agents. The others wanted to make the physicians agents of hygiene, because they themselves were not numerous enough to be everywhere at once. The *Concours* certainly saw them coming. An anonymous journalist writes of "the physicians, whom certain people have the audacity to make the servile, unpaid agents of laws of social protection, passed by the country's representatives without spending a sou" (1900, p. 97). In order to construct their sanitization, the hygienists needed the declining physicians as much as the rising Pasteurians. If they failed to forget this double association, they would never realize their plans.

There is no better evidence of the physicians' sense of being acted upon than this fable written by a physician. On the one side, are sick men, and on the other are the gods, that is, the big bosses in Paris—hygienists, politicians, and Pasteurians: "In the middle are a group of unfortunate beings, for whom there is neither rest nor respite; entrusted with the task of caring for the humans and of warning the gods, they have no other reward for this task than that of avoiding divine punishment; they find themselves caught between the anger of the gods, who accuse them of being too slow, and the hatred of the humans, who regard them as responsible for their misfortunes" (Hervouest: 1894, p. 26). An acted-upon group may either resist by inertia or, if other social groups heed it, identify itself with the wishes of the other groups and switch over to the offensive, itself proposing a deal. The idea of a deal made with the state, that is, with Paris, pops up more and more frequently in the *Concours*.[8] It is possible, the physicians write, that they could agree to carry out all the new things that the state is asking of them and which they refuse with such ill grace, but only in exchange for a suitable reward and above all in exchange for a strengthened defense of the medical profession: "Public opinion seems little disposed to take sufficient account of the services rendered by the physicians, which, nevertheless, everybody demands of them as insistently as ever" (Anon.: 1887, p. 490).

The deal that was beginning to emerge was that the physicians would serve the state, but the state would then rid the physicians of their traditional enemies. As readers of *The Parasite,* physicians would have to say: we will help the state to rid France of parasites, but the state must get rid of those who are sucking our blood—the pharmacists, the charlatans, the nuns, and so forth. (Serres: 1980/1982). Physicians were neither for nor against "science" as such. Nothing ever comes from outside; a group must always re-engender the outside from within its own interests and wishes, that is, translate it. The physicians selected from Pasteurism only those bits that allowed them to strengthen this new deal. One physician exclaims: "Is it not deplorable and revolting, after the wonderful conquests of medical and surgical science, to witness the truly terrifying spread of the illegal exercise of charlatanism in all its forms" (Lasalle: 1888, p. 562). This physician is ready to admire science only in order to crush the charlatans. We should not blame him for this narrowness of vision, for the hygienists did the same. Only the direction of their movements was in no way opposed to that of the Pasteurians. It seems to us more "enlightened" and more "mature" because it assisted the victors, or at least those who appointed themselves as victors.

Even if the idea was emerging of a deal that would make it possible to divert certain forces hostile to the physicians by using them to eradicate other enemies, the physicians would never have made use of Pasteurism if, by an unexpected drift, the Institut had not come within their reach. The vaccine, which was preventive, rubbed physicians the wrong way, since it deprived them of patients who could pay. The serum, invented by Roux and his colleagues, was on the contrary a therapy that was used *after* a patient had been diagnosed sick. Now the doctors, after many other groups, by giving Pasteurism a push would also advance their own interests. The *Concours Médical* allows us to date, week by week, the movement by which a group, hitherto acted upon, switched over to action because the others had moved in other directions. The Pasteurian shift from vaccines to serums via immunology, provided the physicians from 1894 onward with a way of continuing their traditional profession as men who treated patients with an efficacy *reinforced* by Pasteurism.[9] At the cost of a little laboratory equipment, they gained the means of diagnosing and treating diphtheria, a terrible childhood disease. The Pasteurians then offered physicians the equivalent of variation in virulence, which the hygienists had used immediately in order to translate it into "contagion environment." As soon as they were able to go on doing what they had been doing, the same physicians who had been called narrow and incompetent immediately got moving, an exemplary proof of the falseness of the diffusionist model.

The reversal of attitudes may be summed up in two sentences. One is Richet's: "We must insist that the physician make use of these methods of serotherapy" (1894, p. 417). This was the position of the groups that had become dominant, which had had the initiative for twenty years and wanted to drive the physicians into reforming themselves. In the *Concours* a week before we find: "So let us not enthuse too quickly lest we subject M. Roux's discovery to the fate of M. Koch's on tuberculin and examine the facts thoroughly; above all, let us convert our clients to our skepticism and not let ourselves be influenced too quickly by ideas, which they appear to have adopted uncritically from their newspapers" (Anon.: 1894, p. 434). Two definitions of the role of physicians, science, and the public are in conflict here, and the two sides are lined up with the *Revue* and the *Concours*. What is at stake is simple enough. If the public raises a hue and cry for the serum from the Institut Pasteur that may save its children, what are physicians to do? Reform at last and *give into pressure,* says the *Revue;* remain skeptical and *resist pressure,* says the *Concours*. This is the collision point of two immense forces. The physicians should give in and become at last the modern agents that we need; the physicians should resist and continue to keep the public away from these somewhat unscientific enthusiasms. But the physicians were neither to give in nor to resist; they were to deflect their course.

In October 1894 the big story in the *Concours* is not diphtheria but the condemnation by the courts of a physician, Dr. Lafitte. Roux's serum is mentioned by the physician writer, but still with a view of counseling scientific prudence: "M. Roux's discovery continues to raise a unanimous movement of enthusiasm. We are happy to observe this and associate ourselves with it. However we cannot but feel a certain apprehension when confronted by 'universal enthusiasm.'" The anonymous writer adds, "We must show the world that the harebrained French are capable, in the sciences, of proving even more cautious than the ponderous Germans themselves!" (1894, p. 510). This sentence was written during the *Figaro* subscription, at the very time when the diphtheria service, which Richet wanted to force the physicians to use, was being set up!

Who can still speak of "dazzling" and "indisputable" proof? This prudence in the face of so much enthusiasm provides a splendid contrast with the tendency of the hygienists to extrapolate Pasteur's conclusions even before he had opened his mouth. But the physicians' mistrust is understandable. Let us not forget that credulity, trust, skepticism, indifference, and opposition refer not to mental attitudes or virtues but to an angle of displacement. The same physican journalist explains perfectly why there is so much distrust. In order to diagnose diphtheria with "certainty" and to treat it effectively, a physician has to go physically to the Institut Pasteur twice: the first time to bring in the membranes from the patient's throat; the second time, if the diagnosis of diphtheria had been confirmed by laboratory test, to take the serum vial back to the patient. There is nothing surprising in this, since it was only in the laboratory that the power of the microbes was reversed. In order to move the bacillus, the physician had to move himself twice in the direction of the Institut

Incomprehensible Revolution   *Sudden Conversion* [handwritten annotations]

laboratory; that is, he twice had to deny the local work of the practitioner. "This system is absolutely impossible," the physician writer adds. Physicians could translate the diphtheria serum only if it was moved to them and enabled them, by this new means, to do better what they had been doing before. If it was a matter of going twice to Paris and thus reinforcing the Institut Pasteur, they would not move. What could be more natural? This is not slowness but *negotiation*.

In January 1895 the physicians' resistance was weaker. They were no longer complaining of haste and universal enthusiasm, but of bad organization in the serum department. Why? Because that organization was set up with the express aim of *moving the serum at last to the physicians' consulting rooms*. The scope of this movement was in direct ratio with the decline in their mistrust. In the negotiations that were taking place, any diffusion of the serum to the consulting rooms reinforced at last their position as traditional practitioners capable of diagnosing and treating a disease. The Pasteurian route no longer interrupted or ridiculed their work but, having itself been deflected, gave comfort to the physicians. The price to be paid was fairly low; all it required was that the physician's consulting room should be transformed *at certain points* into an annex of the laboratories of the Institut Pasteur. Physicians needed to learn the use of a microscope, a few techniques of culture, and a few gestures to inoculate the serum. This continuous displacement in the point of application of the forces would become so wide that the physicians ended up aligned more or less in the same direction as the Pasteurians, who had themselves deflected their researchers from vaccines to serums. If we had not carefully reconstructed these two displacements, it would have been inevitable to speak on the one hand of an incomprehensible "revolution" and, on the other, of a sudden "conversion."

Let Us Prepare for Evolution If We Are To Avoid a Revolution

The breaking point came in the *Concours Médical* on March 23, 1895. Jeanne, a future editor of the journal, proposed to his colleagues that they move around 180 degrees. He wanted to switch from the defensive to the offensive. This article, astonishingly entitled "Bacteriology and the Medical Profession," is quoted here in full:

It may be not too soon to look ahead into the future that the scientific revolution, brought about by the beneficent discoveries of the illustrious Pasteur and his school, has in store for the medical profession.

What a distance has been covered since the deafening duel between those two orators in the Académie, Pasteur and Peter! And yet it seems only like yesterday. The ardor and skill of the champion of our old clinical methods were wasted, for the adversary advancing against him was not a theoretician, one of those dreamers who create a fashion, a passing fad, but it was a scientist, it was the experimental method, it was progress.

So today his army holds all the keys of the fortress.

Surgery and hygiene have been conquered: the old medicine is no longer able to fight alone for the terrain. Diagnosis, that primordial element of our art, will soon no longer be able to do without the microscope, bacteriological or chemical analysis, cultures, inoculations, in a word everything that may give our clinical judgments absolutely precise data.

But what will then become of medical flair, that indefinable something that we believe to be ours, and experience, that guarantee which the public used to require of our white hairs? Their value will be disputable and will be more and more disputed.

So it is with anxiety that we consider the future of our country physicians, who left the Ecole de Médicine even ten years ago.

When that flock of practitioners who now *crowd* the faculties spreads out into our provinces and, by that very fact, makes us tremble before such stiff competition; when the struggle for existence begins, between us and those young men armed with different skills from ours, with the ardor and confidence that a sense of real value gives one, will we not soon be threatened with a crushing, irremediable defeat? Will the public be on our side?

Colleagues, forgive us for this cry of alarm!

From the heights of our settled situations, we should no longer laugh at bacilli and culture media. Those who cultivate them already deserve our respect for the services that they have given mankind; for us, the old guard of the medical profession, they must also inspire salutary fear and a determination to be useful. We must march with the times. The coming century will see the blossoming of a new medicine: let us devote what is left of this century to studying it.

Let us go back to school, and prepare the ground for an evolution, if we are to avoid a revolution.

Precious victories of science Available to all

And if it is impossible for many of us to leave our native soil for the lecture halls and laboratories of our young masters, let us seek their teaching where it is to be found, that is, in the medical journals. In our day, treatises and dictionaries are out-of-date even before they appear. Only the journal can follow the rapid march of progress and scientific evolution. Let us read more.

In this way we shall take possession of the theory of the new ideas. Then, throwing off ill-placed smugness, guided solely by good faith and a love of truth, we shall ask our young compet-itors, at the patient's bedside or in consultations, to share the benefits of their recent studies with us; at the same time let us tell them that, by way of compensation, we shall share with them the fruits of our experience in the skills of the medical profession. Service for service. In this way we shall establish and make closer the bonds of professional solidarity, which will thus make the precious victories of science available to us all. (Jeanne: 1895, p. 133)

This Dr. Jeanne is like Prince Salina in *The Leopard* faced with revolution (Lampedusa: 1960). If he switches over to the offensive, it is to keep everything the same and to bar the way both to colleagues and to the enemies of medicine alike. It is Jeanne and not I who is stressing the balance of forces, who develops the military references, and who speaks of contract and cooperation only in order to escape from a desperate situation. In passing over into action, those who were previously seen as inactive were obviously to betray what was expected of them. They were to shift the function that had been given them and which they now seized avidly. They accepted the role offered them that they had once stubbornly refused. "In our time social life tends, on the contrary, to use medical knowledge more and more. Governments, courts, authorities of all kinds constantly call upon our technical competence. There is nothing in this testimony of esteem for our skills and professional courage to displease us. Let us, then, accept them with good grace. But let us not lose the opportunity of reminding those authorities that all work deserves its reward" (Jeanne: 1895, p. 144). The physicians had stopped dragging their feet, but they were now asking for payment that the others were not willing to give them. Having been moved, they now agreed to move of their own accord, but only in exchange for something else and to go where new and better-paid work awaited them. Either the groups were not interested and nothing would make them change their mind, or they were interested, but only in translating in a different way what they had understood.

Nothing could better show the complete change of attitude than the position of the *Concours* toward Pasteurian science: "Until recent years, the *Concours Médical* has voluntarily abstained from speaking of microbes, rodlike cells, comma bacilli, and cocci (strepto, staphylo, mico, etc.), or of pure bacteriological studies, knowing that practitioners, its usual readers, would not care very much for that overspeculative, overhypothetical hodge-podge." They *voluntarily* maintained their distance from the Pasteurians. What is the point of learning what cannot be translated? What is the point of believing in something that offers nothing in exchange? What is the point of giving credence to what encourages the spread of enemies? In 1895 everything changed when it became possible to see diphtheria as a way of saving traditional medicine. As Jeanne recalls: "But today, bacteriology has emerged from the laboratory, it has entered clinical medicine, it has even reached therapeutics." It is not I who am speaking of displacement. It is Jeanne who gauges the movement of the Pasteurian laboratory, which finds itself at last in a place where it can serve the physician: "From the beginning it declared its superiority; the whole of France already possesses a very powerful serum against diphtheria." He adds this final blow: "It is absolutely necessary for every practitioner to know this treatment and to be able to apply it. It is about time for everybody to be aware of this enormous progress" (Jeanne: 1895, p. 199). What happened to the prudence invoked in September? What happened to the need to appear "more cautious than the ponderous Germans themselves?" The movements of diphtheria have altered the direction of that "absolutely necessary." Yet they were the *same* practitioners.

In April they went further still. The *Concours* demanded as a right that the physician go back to school and learn bacteriology: "Just as the new laws make all physicians official agents of the public hygiene service, those agents must be provided with the means of learning and playing their roles" (Anon.: 1895, p. 160). A fine word, that of "agents." As patients, physicians mocked the little beasts; as agents, they wanted to know everything about them. The contract had changed its meaning. The country was right to demand that physicians learn the new sciences. But now it was a right that physicians demanded in exchange for what the country demanded of them.

Conquering Our Conqueror and Translating Our Translators

In September 1895, just a year after the celebrated skeptical editorial, we can read in the *Concours:* "The physician who deprives himself of microbic control in cases of exudate [that is, the inspection of patients' throats] would be as irresponsible, heartless, and guilty as the doctor who, in the case of pulmonary disease, refrained from using auscultation" (Anon.: 1895, p. 383)

After such evidence can it still be said that "time passes" or that there is a time that serves as a frame of reference for history? It was only now, fifteen years after Pouilly-le-Fort, that physicians were realizing Pasteurian bacteriology had emerged from the laboratory. They were therefore desperately slow. Yet within a year the same "narrow" and "limited" physicians had overcome their scruples, so that it was now criminal not to do what it would have been dangerous to do the year before. Physicians were therefore moving at astonishing speed. But in fact they were moving neither slowly nor quickly, since in 1895 they were transforming antidiphtheria vaccination into something as venerable, as *traditional,* as obvious as the auscultation invented sixty years before. Time is negotiated: that fact is obvious enough, yet obvious as it may be, it is all too often forgotten by historians who explain social movements by one of the ultimate and distant consequences of those movements, namely their position along the arbitrary grid of days, months, and years.

Indeed, from the very day that physicians moved into action, they *immediately altered the chronology* so as to include Pasteur as one among other elements of the old, at last triumphant medicine. The rearrangement of the secondary mechanism is nowhere clearer than in an article by Bouchard in the *Revue Scientifique.* It is the first article in that journal in which physicians are now talking like the surgeons, hygienists, and army doctors, that is, proudly and in the first person plural. Of Pasteur he writes: "But whatever the importance of a medical discovery, it does not supersede medicine, it can find its place in it." We are far from Pasteur's takeover of the old medicine. Who has moved? It is Pasteur who is included in medicine, whereas he claimed the contrary. Bouchard goes on: "The contribution of bacteriology is strangely reduced, and for that reason we remain within the old medical doctrine, which many thought we were turning away from." His interpretation of serodiagnosis is the exact equivalent of what the hygienists were saying twenty years before about the contagion en-

Hygiene emerged from the laboratory
& crept out into the world

vironment: "This serotherapy exalting the functions by which we defend ourselves naturally from microbic invasion also finds its place in naturist therapeutics." He ends by consecrating not the rallying of the physicians to Pasteurism but the final absorption of the Pasteurians by the physicians, comforted at last: "Do you not think that this great therapeutic progress, far from shaking the old edifice, usually does no more than solicit the efforts of the old curative nature?" (1895: p. 225). As Héricourt said ten years before: "The old adage *Morborum Causa Externa Morbus Corporis Reactio* is therefore as true as ever" (1885, p. 532).

When at last the physicians switched over to the offensive, they redefined the role and function of those who had hitherto claimed to define them. The acceptance of laboratory methods was renegotiated according to the terms of the old clinical medicine: "Radiography, bacteriology, serodiagnosis are still weapons of too quick a trigger for ordinary mortals, I mean, for practitioners like us. We may dream of the precision that they promise us, but we must not forget that they sometimes have serious drawbacks and we must subject them unflinchingly to pure clinical medicine. Above all let us not start a civil war over germs" (Jeanne: 1900, p. 145).

We see the extent to which Pasteur's takeover of medicine was an illusion. The doctors whom he needed to extend his influence were not as obliging as the hygienists, who elected him to be the leader of their movement so as to make their own conviction efficacious. As late as 1905 the *Concours Médical* claims with a certain superiority: "Our readers are not unaware that we have always been among those who claim for clinical medicine a formal right of priority over the laboratory and bacteriology, and of course in our period, which is so enthusiastic about laboratory methods, we have not decided to change our opinion. But, though of secondary importance, the diagnostic method provided by the laboratory must not be disdained" (Huguenin: 1905, p. 202).

A generation after the enthusiasm of the hygienists, the attitude of physicians was simply not to disdain the laboratory. Even bacteriological science had been completely retranslated.[10] In 1900 the *Concours Médical* launched a competition among its readers to propose remedies for professional *overcrowding*, the only truly fatal disease from the point of view of the physicians. The prize went to a certain Dr. Gouffier (1900, pp. 528–556). He spoke of the bacteriological sciences of the great Pasteur, but he saw them as a *remedy* for overcrowding! By extending the period of medical study, the sciences would limit the number of colleagues, and if Greek and Latin were added, the results would be better still. This was no act of meanness on the part of the journal. In 1906 the *Revue Scientifique* also launched a questionnaire on the "Reform of Medical Studies." Toulouse, the new editor of the *Revue,* began this inquiry because, he explains, "a considerable movement is emerging whose aim is the abolition of the so-called secondary sciences (chemistry, biology, physics, parasitology) in medical teaching and an orientation toward the training of experienced practitioners with a thorough grasp of the practices of their art" (1905, p. 702).

Once again time does not move in one direction. There are as many directions as there are agents capable of making their positions irreversible. It was possible to get rid of bacteriology as quickly as Pasteur got rid of medicine. Both movements were possible on condition that allies were found. It was possible to move from vaccination to the absence of vaccination, just as to move from scientific medicine to a medicine without science.

When the physicians switched over to the offensive, they certainly took something from Pasteurism, but unlike the hygienists, they did not take the laboratories; they took the prestige attached to Pasteur. The notion of legitimacy is rarely correct in sociology, but it may sometimes serve to designate the relation between groups that cannot translate their interests in order to extend their influence but that nevertheless need one another. It is another, simpler form of association. A general inventory of fees in 1905 for all medical treatment, which the *Concours* published in order to ease out competition, shows fairly clearly the place played by Pasteur's science in basic medicine. For a simple aseptic bandage the patient will pay the price of a consultation. This was Pasteurism of thirty years earlier. The price of five visits or consultations includes "subcutaneous injection of antimicrobic and antitoxic serums, including the treatment of local preventive accidents." After fifty years of laboratory work and thirty years of resounding declarations about the disappearance of infectious diseases and the establishment of the new medical science, only a few lines have been added to pages and pages of what was done before. The radical epistemological break turns out to be a thin scratch in the practice of the majority.

But this was not true of the prestige gained by this strategic reversal.[11] The physicians, whom the hygienists needed so much in order to guarantee medical supervision, were converted at last to this role and ended up occupying the terrain that others had left them because they had failed to create a new professional group. Those who say of the physicians that they were both contemptible and indispensable wanted them to become both scientists and policemen. The physicians accepted the role of policemen, abandoned that of scientists, except to extend the duration of studies and to adopt a few methods that were "not to be disdained," and they retained the prestige of the agents who needed them. Others acted as their cat's paw.

During the rest of the period, innumerable articles expressing self-satisfaction were published in the *Revue Scientifique*. Nothing had really changed in style or technique. Only self-confidence had become greater. Landouzy, the great mixer, writes in his inimitable style about the physician: "Less absorbed in the assistance to be given to patients, the physician will now devote himself to an enterprise that has scarcely begun, that of increasing the vitality of the individual and the species . . . Child-rearing, breeding, 'hominiculture'; family hygiene, educational, psychological, and physical hygiene. Is it nothing to teach and to practice all that?" (1909, p. 162). Indeed it was not nothing! The totality sought by the hygienists was inherited by the physicians, since there was no specialized profession in public medicine. The Pasteurian spring itself was *imitated* by the physicians, who spoke of everything, but without adding the laboratory at strategic points. In 1914 Chauffart, in an article entitled "War and the Health of the Race," defines the extension of medicine:

> Over the past fifty years, our medical habits have changed remarkably. Once, as physicians, we saw scarcely anything but the *individual* to be treated; we were the physician of a patient, we tried to treat him, to cure him, and we thought that once he was cured, we had completed our task. Then, as the general orientation of ideas altered, we saw that the physician had a place not only at the patients' bedside, but also in advising a family, in the counsels of society, the state, and that, in fact, just as there are individual healths, there are healths of the nation and healths of the race, and the physician must concern himself with these just as much. (1915, p. 18)

The physician, too, had moved. He moved from the patient to the state. He occupied the place appointed by the others and changed his habits as little as possible. Truly a strange revolution, in which the supposedly revolutionized groups moved only when they were sure of being able to carry on as before and betray their translators. The revolutionary history of the sciences still awaits its Tocqueville.

Coercion at Last

Between 1871 and 1919 the various actors tried through every possible relationship to define hygiene, science, and medicine in relation to one another. A few articles lay claim to clarity and wish to be purely medical, purely scientific, or purely hygienist. But most of them exploit an ambiguity to fuse interests and to produce new mixtures. Richet exclaims: "All the problems of hygiene are social questions. Now what could resolve them if not the Sciences?" (1888, p. 360). All the attempts to classify or distinguish these entities have no interest in themselves; they merely indicate the number of freeways that were converging on this enormous cloverleaf exchange: the social question (poverty, exploitation, alcoholism, tuberculosis); the falling birthrate and the physical weakness of the French (gymnastics, army, wet-nursing); questions of sanitation (drains, pure water, pollution); the link between diseases and large-scale international commerce (quarantine, supervision); surgery and hospital administration; the resistance of bodies and immunology; infectious diseases; and parasitical or tropical diseases. There was no one word to cover all these concerns. Moreover, the *fusion* that enabled the Pasteurians to move from one order of concern to another could not last very long once the actors had mixed together, moved position, and reached their "aims"—or what they had decided to call their aims.

Hygiene, for instance, that translator of such an important social movement, gradually disappears from the *Revue*. Like the microbe itself, it was an agent endowed with unity and personality for only a short period. There was still talk of hygiene, but it was no longer that "sender" in whose name one had to act. Articles in the review no longer advocate changes; they now *describe* organizations, such as "the Paris sanitary organization" and "the present state of methods for purifying water." In these descriptive articles the words "supervision," "regulations," and "policing" recur constantly. In 1910 a report of the Académie de Médecine on typhoid fever proposes "supervising" the catchment of springs, "supervising" the purifying machinery, "regulating" sewage and "detecting" contagious diseases; "the prefectorial authority has a duty to make sure that the said regulations are carried out." What we now have is *routine,* since it is no longer a question of setting up or extending networks by using elements from elsewhere to maintain them. The laboratory, continues the Académie, is a "valuable help," but it is no longer anything more than the Bertillon of hygienic policing, "by supervising the healthiness of the water supply, by helping the physicians to arrive at a diagnosis. It is desirable to set up bacteriological stations in the departments; without such stations the sanitary policing of the municipality and departments cannot be carried out effectively" (Anon.: 1910, p. 471). The laboratory was no longer at the forefront of the struggle in which the society of the time was engaged; it was now no more than the informer of an administration that had spread everywhere, that had conquered the microbes and the public authorities, and whose laws now had only to be *implemented.*[12]

At the time when hygiene was disappearing as an actor, to be gradually replaced by medicine and the physicians, when all the great programs of sanitation were finally underway, and when the law of 1902 had been passed, there was no longer any controversy.[13] No one was trying any more to win allies for the spread of hygienic precepts. Scientific laws ratified by juridicial laws no longer left any room for argument among the groups already recruited. Or rather, as the hygienists' allies became more numerous and more highly placed, the remaining enemies could be destroyed with less compunction. In 1887 Rochard wants hygiene to be militant, but he also knows that it has to *negotiate:* "If hygiene wants to have the last word and to get its decisions respected, it will do well to move with extreme moderation and caution. If it shows itself to be tyrannical, interfering, intransigent, it will inevitably fail. It must be a protection, not a fetter. It must impede the workings of the great economic cogs of the country only when absolutely necessary" (1887, p. 392).

A few years later Armaingaud, speaking on the subject of tuberculosis, uses both a model of translation and a model of diffusion. He complains that ideas about the contagion of tuberculosis do not spread, but at the same time he looks for allies powerful enough to force the diffusion of a practice. He proposes to put up notices requesting industrialists to take care of their tubercular workers: "Do you believe that it will be possible to overcome all the obstacles put up to oppose these measures by the apparent or real interests of factory and workshop managers ... without a national campaign to shift public opinion and force their hands?" This illusion denounced by

Armaingaud was nevertheless shared by all those who believed that Pasteur's discoveries moved through society of their own accord. As a good propagandist, Armaingaud knows very well what forces must accompany an idea if it is to be capable of undertaking the journey: "Once they have been enlightened and convinced, the industrialists must, if they are to make up their minds to act, be forced, with few exceptions, by the demands of the interested parties." Armaingaud is under no illusions. He wants to stir up the workers against their bosses so as to convince the bosses that Koch's bacillus is dangerous. This move is of course in the best interests of the industrialists themselves, but if they are to understand these interests, they need *strong* proof: "If we are beginning to get the disinfection of hotel rooms implemented ... it is thanks to the publicity already given by the newspapers to the contagiousness of tuberculosis, which is making new arrivals demand guarantees" (1893, p. 34).

What a good *associo*logist he is! He knows that a proof proves nothing of itself and that the social body has to be worked upon if the press is to influence customers to influence hotel-keepers to influence disinfection services to drive Koch's bacillus out of society. He sets the actors against one another, knowing that only a slight displacement can be obtained from each of them and that a mere "diffusion" of the "truth" is not to be expected.

Yet a few years later everything changed once this series of people had been convinced. There was no longer any need to negotiate, to recruit, or to set different forces against one another. There was nothing but inert bodies: "The prevention of disease is undermined by the public's ignorance and carelessness and by the irrational resistance that it sets up against hygienic measures" (Arloing: 1910, p. 481). There they were at last, the "inert," the "irrationalists," and the "resisters." No one had to be handled carefully any more. The angle of interest was not sharp enough to make it worth the trouble to return these people into accomplices. The role of disinfected expected of them was not complicated enough to need to seek their active connivance. *Those who took forty years to become convinced that Pasteur was interesting* and who took the trouble to understand him only when they were *sure that they would be able to carry on the same activity* regarded the slowness of other agents to cooperate with them only as *inertia*. The story is finished; there is no longer any subtle analysis to be made. The poor were the ones who were now besieged by the hygienists, the biologists, the public authorities, the physicians, the surgeons, the midwives, the prefects, the mayors, the disinfection services, the teachers, the army doctors. Each of these groups had discussed, negotiated, diverted, resisted, distorted the struggle against the microbes. But once allied to one another, clinging to all the measures that they had taken to make their positions irreversible, they became indisputable. No negotiation. No distortion. Implementation.

Hygiene has often been discussed in terms of policing and coercion. Though probably correct, these terms only describe its final state[13]. Those who speak of "disciplining" and of "domination" can begin their analysis only at the point where almost everything is over, where power has no longer to be "negotiated," where all that remains is to convince those at the bottom of the ladder. In the 1910s, indeed, the triumphant hygienist movement must have seemed like a disciplinary authority. But to limit the analysis to this coercion is to understand nothing that happened before, when hygiene was weak, voiceless, and powerless but aspired to power. Once it had become allied with the forces that mattered, it was then possible to follow its progress as it battened on the poor, delousing, reclaiming, vaccinating, and washing them. There is no shortage of sociologists to do that. They think that they are denouncing power and ignore the decades during which the hygienist movement was trying to claim power, without having it, by looking in such unexpected places as the laboratory.

Portrait of the Pasteurians as Solon of the Tropics

While hygiene was being incorporated into a vast bureaucracy and having no other problems than that of being implemented, while the physicians were occupying the terrain of both the hygienists and the new medical sciences, without having to alter anything except their prestige and the new role of sanitary policing, whose allocation to them they finally accepted, while the Institut Pasteur, making vast strides in immunology and biochemistry, was moving away from the point of fusion won earlier by Pasteur and seeming more insulated in its laboratories, a new movement on the part of the Pasteurians was beginning to restore to them the central role that they had had in the redefinition of society during the 1880s and 1890s.

To follow this transformation of a society by a "science," we must look not in the home country but in the colonies. The enormous part played by tropical medicine in the production of the *Annales de l'Institut Pasteur* revealed most directly the struggle between micropar-

asites and macroparasites, and it was there that the forces thrown
into the balance of the Pasteurians could tip the scales irreversibly in
favor of the westerners. It was in the tropics that we can imagine best
what a pasteurized medicine and society are: "It is clear that even
more than the heat, which is at most an unpleasant factor, fever and
dysentery are the "generals" that defend hot countries against our
incursions and prevent us from replacing the aborigines that we have
to make use of " (Brault: 1908, p. 402).[14]

The blacks, like the Hovas, were immunized. The westerners were
not. Thus the natives had a superiority that compensated for their
natural inferiority. It was therefore necessary to reverse once more
the balance of forces and to restore to the westerners their natural
superiority, by overcoming that relative ally of blacks and that enemy
of whites: the parasite. Calmette writes: "Is it unlikely that Africa
would have aroused so much greed if the peoples of Europe who now
share it had not been counting on their victory over malaria" (1905,
p. 417).

To situate the Institut Pasteur in this gigantic struggle, we do not
even have to be crude Marxists or to resort to far-fetched evidence.
In an article entitled "The Scientific Mission of the Institut Pasteur
and the Colonial Expansion of France," Calmette writes again: "It is
now the turn of the scientific explorers to come onto the stage . . . their
task is to draw up inventories of the natural resources of the conquered
countries and to prepare the way for their exploitation. These scientific
explorers are the geographers, engineers, and naturalists. Among the
last, the microbiologists have a considerable role to play in protecting
the colonies, their native collaborators, and their domestic animals
against their most fearsome, because invisible, enemies" (1912, p.
129).

This work on the parasites had a direct influence on colonization,
because parasites directly limited the extent of the empires formed by
the macroparasites (McNeill: 1976). The identification and movement
of each parasite made it possible to advance further. The extent of
this shift in favor of the whites can be seen quite clearly. It is one of
those dramatic proofs beloved of so many scientists. With each parasite conquered, the columns of soldiers, missionaries, and colonists
became visible on the map of Africa and Asia, sailing up the rivers
and invading the plains, just as, thirty years before, the surgeons
tackled new organs with each step in the progress of asepsis: "So it
is thanks to these two scientists (Bouet and Roubaud) that we now

know the various modes of propagation of the trypanosomiases that form the principal obstacle—one might almost say the only obstacle—to the exploitation of the enormous African quadrilateral that extends between Guinea, the Upper Nile, Rhodesia, and Angola" (Calmette: 1912, p. 132).

This politicomilitary role given to the biologist was explicitly claimed by the Pasteurians. Roux, praising the work of Laveran in 1915, exclaims: "Thanks to them (the scientists), lands that malaria forbade to the Europeans are opened up to civilization. It is thus that the work of a scientist may have consequences for mankind that go well beyond those of the conceptions of our greatest statesmen" (1915, p. 410). Yes, that's it: they go *beyond* those of the greatest statesmen, because instead of pursuing politics with politics, the scientists were pursuing it with *other means*. This unforeseeable supplement of force gave them that superlative politics which made it possible to act on the poor, on the inhabitants of Madagascar, on the surgeons, on the Africans, and on the dairies.

Pasteur was hailed as a more famous conqueror than that of Austerlitz. Nevertheless, when he put up for the Senate, this great politician was beaten hollow. This outcome says everything. Political politics fails, but politics by other means succeeds superlatively. Invade Africa with a determination to dominate with power, and you will be dead before long and be confined to the coast. But invade it with the Institut Pasteur, and you might really dominate it. What was unforeseeable was that the fusion of the Pasteurian laboratory, tropical medicine, and tropical society would be much more complete than in France itself.

This new medicine had five characteristics that explained its success. To begin with most of the diseases themselves were new. Clinical medicine was either nonexistent or being practiced only by army doctors, the first to become pasteurized. The Pasteurians, then, did not have to deal delicately with a century-old clinical medicine.

Secondly, the diseases that could be studied were all derived from germs or parasites. The other diseases, which in France itself made up nine-tenths of the work of the medical profession, were simply ignored. Among the colonists the potential patients were all young men in good health who fell ill from infectious diseases. When doctors treated the natives, they did so en masse, working on devastating symptoms and spectacular diseases (plague, yellow fever, leprosy, sleeping sickness). In such a situation there could be no question of a family medicine, in which the patient was expected to pay.

Third, most of the diseases were connected to the life style of insects. No physician was prepared, by training, to be concerned with entomology. The Pasteurians, busy crossing the frontiers between the different sciences, could, without requiring too much retraining, add new species of actors to the swarm of microbian agents: all the great discoveries of this period consisted indeed in rediscovering the route by which a parasite, an insect, and a man were linked. The talk was now entirely of fleas, mosquitos, tsetse flies, and parasitology expanded: those insects were themselves subject to parasites that used them in order to move or reproduce. It was decidedly a great period, for the readers of Serres (1980/1982) and the so-called social actors seemed to rival in ingenuity the so-called "natural" actors in learning how to overlap, move, and become contaminated.[15]

The fourth characteristic of this new medicine was that there was no other medical corps on the spot, except the witch doctors, who were already fighting on a different terrain. Nothing was there to force the Pasteurians, often army doctors by training, to limit the scope of their activities. Whereas at home they were preceded by innumerable professional groups interested in health and were ultimately swamped by them, in the colonies they could construct public health from scratch. This is not a metaphor. They often preceded the towns, which they could therefore build according to the strictest recommendations of hygiene. At home they had always to take into account centuries of insanitariness and the doubts of public authorities. In the tropics the secular arm of the military authorities was on their side. If all the houses had to be rebuilt, then they could be.

The fifth reason for the success of the new medicine was more paradoxical. The Pasteurian spring consisted in the culture of a microorganism and its attenuation, then in the manufacture of attenuated microbes or serums. Now the parasites were giants compared with the microbes and did not allow themselves to be grown, let alone attenuated or inoculated. This failure, due to a new ruse on the part of the diseases might have cut short all the efforts of the Pasteurians. Instead, those efforts were shifted. Since the Pasteurians could not concentrate all their attention on the laboratory stage and could interrupt the parasite only by interrupting his life cycle in life-sized conditions, they had to obtain plenary powers and always act on a large scale. Since they could not reduce their contribution to one stage and leave others to apply it, the Pasteurians had to be allowed to *legislate* for the entire social body.[16] Malaria or yellow fever were to be destroyed not with vaccines but by ordering the colonists and natives to build their houses differently, to dry up stagnant ponds, to build walls of different materials, or to alter their daily habits. The Pasteurian worked both in the laboratory and on administrative regulations, but his actions could no longer be studied in distinct stages. He legislated like Solon: "It has taken thirty years for science to discover the nature and origin of all the great endemic diseases that seemed to have stopped civilization at the threshold of the tropical countries. All the problems have now been posed, all the solutions are in sight. The governors of our colonies think as men of science and act as administrators to apply the doctrines to which the century of Pasteur has given birth. Our corps of colonial health is continuing with its admirable work everywhere" (Nattan-Larrier: 1915, p. 303).

The means by which administrators were enabled to act as men of science was, as always, the laboratories, now extended to all the colonies, at Saigon, Algiers, Tunis, Tangiers, Brazzaville, Dakar. In 1901 an Institute of Colonial Medicine was founded by a subscription of the Union Coloniale Française and was attached to the Institut Pasteur. In 1908 the *Bulletin de la Société de Pathologie Exotique* was added to the *Annales:* "In the Far East and in French Africa, then, there is no longer any one of our colonies that has not possessed for several years one or more laboratories suitably equipped for bacteriological research and for the immediate application of Pasteurian methods either to the treatment and prevention of infectious diseases or to the study of the economic conditions dependent upon biology" (Calmette: 1912, p. 133).

The role of both preventive medicine and the rise of the standard of living in the decline of the great infectious diseases in Europe has been a matter of dispute. But there has never been any doubt as to the direct and determining role of the Institut Pasteur in colonization. If it had been necessary to make colonial society only with masters and slaves, there would never have been any colonial society. It had to be made with microbes, together with the swarming of insects and parasites that they transported. It is not enough to speak shyly of the "influence of parasitology on social or institutional interests" (Stepan: 1978). With only whites and blacks, with only miasmic regions and healthy or dangerous climates, that Colonial Leviathan which spread across the globe could never have been built. Nor can the colonial medicine of the Pasteurians be explained in terms of "society" and its "interests," since the Pasteurians were capable, once more, of moving their programs of research sufficiently to obtain a richer definition of society than had all the exploiters or exploited of the period. The Pasteurians reshuffled the cards by daring to change profoundly the list of actors playing a role in the world, by modifying the trials of strength, and by inserting the laboratory into the strangest and least predictable place. Their "genius" lay in that they twice succeeded, in two different periods and two successive political situations—first at home on infectious diseases during the 1880s and 1890s, then in the colonies on the parasitical diseases before 1914—to reorder society in a way that went well beyond the "conceptions of our great statesmen."

Chapter 4
Transition

In this part of the book I chose an indisputable, revolutionary, esoteric science, whose applications alone, outside the laboratory, had a prodigious influence on various groups—some open and modern, which adapted to them; others closed and backward-looking, which remained inert. Before such a succession of mysteries—the mystery of the invention of facts, the mystery of their diffusion, the mystery of *adequatio rei et intellectus,* the mystery of recognition—it was possible to challenge the agnostic (an incompetent sociologist) to provide even the beginning of an explanation. All one could do was to keep silent, to be content, to admire theories, to write glosses on the "social," or worse still, to study nothing but the "symbolic and cultural dimension," the bone that those who have given up the good fare of reality are content to gnaw.

By means of this journey through the weakened microbes I think I have shown that this vision of the sciences and of society is a myth, our myth, the only one to which we who think ourselves so clever subscribe in simple faith. The sciences have no more content than the social groups. Those two symmetric phantasmagoric beings are obtained only by a reductio ad absurdum, and we are only just beginning to perceive both its danger and how to face up to it.

Out of the magical combat between "Napoleon" and "Kutuzov," Tolstoy created a battle of crowds, which act sometimes in great masses and sometimes as individual characters. In the middle of those crowds, acting sometimes as crowds and sometimes as characters, Napoleon and Kutuzov (without quotation marks), among other things that they do, give orders which are misunderstood, wrongly obeyed, badly transmitted, distorted, and betrayed, and which culminate, from hour to hour, in the movements of regiments and cannon about which the information comes back belated, distorted, and betrayed. The words that the troops give to what is happening also act as self-fulfilling prophecies. Depending on whether someone shouts the word "victory" or "each man for himself," this or that part of the front retreats or rushes forward. The battle to know what is happening and what has happened is endless. The stories begun so warmly between Fabrice and the canteen girl at Waterloo end coldly in the archives and manuals, where they continue to influence the history of Europe and to stir crowds, enthusiasms, and responsibilities. Nowhere can we escape from the consequences of the translations and trials, which are the things themselves. We can never do better. We can never know more clearly.

The same goes for that war and peace of the microbes, which I have recounted so sketchily, as we wait for someone to turn up who will describe the Natasha of rabies and the Prince André of yellow fever. I had to give back to the sciences the crowd of heterogeneous allies which make up their troops and of which they are merely the much-decorated high command whose function is always uncertain. I had to show that these disreputable allies (hygienists, drains, Agar gels, chickens, farms, insects of all kinds) were an integral part of so-called scientific objects. Indeed, I showed that it was in order to make allies of them, to attach to them, to convince them, that they assumed the form of virus, bacterium, or vaccine. Even if at the final vote, the moment of recognition before the tribunal of history, the handing out of medals, those crowds count for nothing, we understand nothing of the solidity of a fact if we do not take into account the unskilled troops. Don't be mistaken. I, too, love the solidity of facts, which is why I cannot be content with those ectoplasms that seem to float around inside scientists' heads. They have no more "content" than they have social "environment." It is during the battle that we redistribute the trials of strength, arbitrarily and temporarily, some as "content" and others as "context." Like the cry of "victory" or "defeat," this is not a description of what the Pasteurians did, but a war cry, to drive back another adversary. It was that war cry which I was supposed to accept as the most illustrious example of reason above the trials of strength, of a real scientific revolution, introduced into hygiene and medicine by Pasteur's theory. And there were people who actually wanted that cry to remain indisputable.

In the beginning, I claimed that I could discuss that indisputable science and provide an explanation of bacteriology because I agreed to recognize it for what it is, a nestled series of reversals in the balance of forces, and because I agreed to follow it wherever it led and to whatever groups it constituted, crossing as often as necessary the sacred boundary between "science" and "society." Have I avoided the three failures that I indicated then: sociological reductionism, the elision of technical content, and the use of tribal words to explain those words themselves? Have I succeeded?

No explanation escapes the trials of strength that I have described—mine no more but also no less than another.[1] What I cannot attribute to the Pasteurians I do not claim to attribute to myself. My proofs are no more irrefutable than theirs, and no less disputable. I must go looking for friends and allies, interest them, draw their attention to what I have written—here extracts from my sources, there cultures placed under their microscopes—and reply in advance to enough objections to convince the reader, so that it will then be more difficult to make a statement as probable as those proposed here. To prove that there are no irrefutable proofs is in no way contradictory.

If readers consider this comparison between the feeble forces of the sociologist and the grand things of which I spoke blasphemous, let them compare the forces, resources, and places to which all these paths lead. There are not two ways of proving and convincing. There is no essential difference between the human or social sciences and the exact or natural sciences, because there is no more science than there is society. I have spoken of the Pasteurians as they spoke of their microbes. We give voice to those whose support is necessary to us. Faithful translators or unfaithful traducers? Nothing is known, only realized through a trial of strength. Politics is probably the best model that we have to understand this relationship between forces and their spokesmen. That which arbitrates in the final resort—fidelity or infidelity, conviction or skepticism—is the *angle* of the direction in which we wish to go. My account will seem convincing only if it allows readers to go faster in the direction that they wanted to go in any case.

In recomposing the forces that made those scientists great and the successive movements that made them admirable, I have not reduced them. On the contrary, I have given them back to those to whom they belong. Where were they going? Pasteur and his followers fought against microbes, made towns habitable, gave the networks of hygienists, surgeons, and army doctors the continuity that they lacked. I like such acts of prowess as much as I like the hardness of facts. I, too, have wept, I admit, when reading their articles, walking around the places that they reached, and seeing the enemies that they weakened. But we are no longer going in the same direction.

It is at this point that the path of the revealers of microbes and the path of people like me part. We no longer have to fight against microbes, but against the misfortunes of reason—and that, too, makes us weep. This is why we need other proofs, other actors, other paths, and is why we challenge those scientists. Because we have other interests and follow other ways, we find the myth of reason and science unacceptable, intolerable, even immoral. We are no longer, alas, at the end of the nineteenth century, the most beautiful of centuries, but at the end of the twentieth, and a major source of pathology and mortality is reason itself—its works, its pomps, and its armaments. This situation was unforeseeable, as was, in 1870, the pullulation of microbes.

Just as the Pasteurians reshuffled the distribution of actors between nature, science, and society by the temporary formation of the microbe-whose-virulence-one-can-vary-in-the-laboratory, we must, to survive, redistribute one more time what belongs to nature, the sciences, and societies. What I once timidly called an "anthropology" or "ethnography" of the sciences has gradually changed its meaning. It first had to study symmetrically all the logical systems, those of the Alladian witch doctors as well as those of the Californian biochemists or the French engineers. But in gradually discovering what made up the logical systems and paths, anthropology finally collapsed. Once the shackles that had paralyzed society and science were broken, we could start to think again about this most ancient object-subject: the world.

Microbes play in my account a more personal role than in so-called scientific histories and a more central role than in the so-called social histories. Indeed, as soon as we stop reducing the sciences to a few authorities that stand in place of them, what reappears is not only the crowds of human beings, as in Tolstoy, but also the "nonhuman,"

eternally banished from the Critique. If we succeed in this emanci-
pation of the nonhumans from the double domination of society and
science, it will be the finest result of that perhaps clumsily begun
"anthropology of the sciences."

However, in order to reach that aim, we have to abandon many
intermediary beliefs: belief in the existence of the modern world, in
the existence of logic, in the power of reason, even in belief itself and
in its distinction from knowledge. I have to write, not as a sociologist
or even as a historian of the sciences, but as a philosopher, and to
define those trials of strength of which I have made such extensive
use in this history of microbes. That is the aim of the second part of
this book.

Part Two

Irreductions

Introduction

Studies about science and society, such as this one on the pasteurization of France, are always met with skepticism. Critics insist that there is something else in science, something that escapes social explanation. After encountering this skepticism for years, I realized that it was not rooted in any lack of empirical studies (though this may play some role) but stemmed from much deeper philosophical arguments about knowledge and power. Knowing that empirical studies would never do more than scratch the surface of beliefs about science, I decided to shift from the empirical and, as Descartes advised us, to spend a few hours a year practicing philosophy. In doing so, I quickly unearthed what appeared to me to be a fundamental presupposition of those who reject "social" explanations of science. This is the assumption that force is different in kind to reason; right can never be reduced to might. All theories of knowledge are based on this postulate. So long as it is maintained, all social studies of science are thought to be *reductionist* and are held to ignore the most important features of science. Although, like the postulates about parallel lines in Euclidean geometry, it seemed absurd to deny this presupposition, I decided to see how knowledge and power would look if no distinction were made between force and reason. Would the sky fall on our heads? Would we find ourselves unable to do justice to science?

153

Would we be condoning immorality? Or would we be led toward an *irreductionist* picture of science and society?

This shift from a reductionist to an irreductionist philosophy closely resembles what happened to Robinson Crusoe when he finally met Friday. I am talking here not about Defoe's story but about the original version of the myth offered to us by Tournier (1967/1972). His story starts off like Defoe's, but halfway through the novel Friday carelessly blows up the powder magazine and Robinson finds himself as naked as he was on his first day on the island. For a moment he thinks of rebuilding his stockade, his rules, and his disciplinary measures. Then he decides to follow Friday and discovers that the latter lives on an entirely different island. Does Friday live like a lazy savage? No, for savagery and laziness exist only by contrast with the order imposed on the island by Crusoe. Crusoe thinks he knows the origin of order: the Bible, timekeeping, discipline, land registers, and account books. But Friday is less certain about what is strong and what is ordered. Crusoe thinks he can distinguish between force and reason. As the only being on his island, he weeps from loneliness, while Friday finds himself among rivals, allies, traitors, friends, confidants, a whole mass of brothers and chums, of whom only one carries the name of man. Crusoe senses only one type of force, whereas Friday has many more up his sleeve. Instead of beginning my philosophical tract with a Copernican revolution—reducing the island to Crusoe's will—I therefore start from Friday's point of view and set things irreduced and free.

For such a view I need, like Friday, no a-priori ideas about what makes a force, for it comes in all shapes and sizes. Some forces are evil and used to be associated with magic and the devil. Others are Aristotelian and seek to realize the shape that lies within them. There are Malthusian or Darwinian forces which always want more of the same and would invade the world with their exponential growth if other equally greedy forces did not check them. There are Newtonian forces which always want the same thing and travel along the same trajectory so long as they are left in peace. There are Freudian forces which do not know what they lust for—displacing, substituting, metamorphosing, or paralyzing themselves as the need arises. There are Nietzschean forces, stubborn yet plastic, wills of power giving shape to themselves. And all of these forces together seek hegemony by increasing, reducing, or assimilating one another. This is why the jungle with its tangle of forces grows across the island.

To follow this argument, we should not decide a-priori what the state of forces will be beforehand or what will count as a force. If the word "force" appears too mechanical or too bellicose, then we can talk of weakness. It is because we ignore what will resist and what will not resist that we have to touch and crumble, grope, caress, and bend, without knowing when what we touch will yield, strengthen, weaken, or uncoil like a spring. But since we all play with different fields of force and weakness, we do not know the state of force, and this ignorance may be the only thing we have in common.

One person, for instance, likes to play with wounds. He excels in following lacerations to the point where they resist and uses catgut under the microscope with all the skill at his command to sew the edges together. Another person likes the ordeal of battle. He never knows beforehand if the front will weaken or give way. He likes to reinforce it at a stroke by dispatching fresh troops. He likes to see his troops melt away before the guns and then see how they regroup in the shelter of a ditch to change their weakness into strength and turn the enemy column into a scattering rabble. This woman likes to study the feelings that she sees on the faces of the children whom she treats. She likes to use a word to soothe worries, a cuddle to settle fears that have gripped a mind. Sometimes the fear is so great that it overwhelms her and sets her pulse racing. She does not know whether she will get angry or hit the child. Then she says a few words that dispel the anguish and turn it into fits of laughter. This is how she gives sense to the words "resist" or "give way." This is the material from which she learns the meaning of the word "reality." Someone else might like to manipulate sentences: mounting words, assembling them, holding them together, watching them acquire meaning from their order or lose meaning because of a misplaced word. This is the material to which she attaches herself, and she likes nothing more than when the words start to knit themselves together so that it is no longer possible to add a word without resistance from all the others. Are words forces? Are they capable of fighting, revolting, betraying, playing, or killing? Yes indeed, like all materials, they may resist or give way. It is materials that divide us, not what we do with them. If you tell me what you feel when you wrestle with them, I will recognize you as an alter ego even if your interests are totally foreign to me.

One person, for example, likes white sauce in the way that the other loves sentences. He likes to watch the mixture of flour and butter changing as milk is carefully added to it. A satisfyingly smooth paste results, which flows in strips and can be poured onto grated cheese to make a sauce. He loves the excitement of judging whether the quantities are just right, whether the time of cooking is correct, whether the gas is properly adjusted. These forces are just as slippery, risky, and important as any others. The next person does not like cooking, which he finds uninteresting. More than anything else he loves to watch the resistance and the fate of cells in Agar gels. He likes the rapid movement when he sows invisible traces with a pipette in the Petri dishes. All his emotions are invested in the future of his colonies of cells. Will they grow? Will they perish? Everything depends on dishes 35 and 12, and his whole career is attached to the few mutants able to resist the dreadful ordeal to which they have been subjected. For him this is "matter," this is where Jacob wrestles with the Angel. Everything else is unreal, since he sees others manipulate matter that he does not feel himself. Another researcher feels happy only when he can transform a perfect machine that seems immutable to everyone else into a disorderly association of forces with which he can play around. The wing of the aircraft is always in front of the aileron, but he renegotiates the obvious and moves the wing to the back. He spends years testing the solidity of the alliances that make his dreams impossible, dissociating allies from each other, one by one, in patience or anger. Another person enjoys only the gentle fear of trying to seduce a woman, the passionate instant between losing face, being slapped, finding himself trapped, or succeeding. He may waste weeks mapping the contours of a way to attain each woman. He prefers not to know what will happen, whether he will come unstuck, climb gently, fall back in good order, or reach the temple of his wishes.

So we do not value the same materials, but we like to do the same things with them—that is, to learn the meaning of strong and weak, real and unreal, associated or dissociated. We argue constantly with one another about the relative importance of these materials, their significance and their order of precedence, but we forget that they are the *same* size and that nothing is more complex, multiple, real, palpable, or interesting than anything else. This materialism will cause the pretty materialisms of the past to fade. With their layers of homogeneous matter and force, those past materialisms were so pure that they became almost immaterial.

No, we do not know what forces there are, nor their balance. We do not want to reduce anything to anything else. We want instead, like Friday, to feel the island and to explore the jungle.

This text follows one path, however bizarre the consequences and contrary to custom. What happens when nothing is reduced to anything else? What happens when we suspend our knowledge of what a force is? What happens when we do not know how their way of relating to one another is changing? What happens when we give up this burden, this passion, this indignation, this obsession, this flame, this fury, this dazzling aim, this excess, this insane desire to reduce everything?

Chapter 1

From Weakness
to Potency

1.1.1 Nothing is, by itself, either reducible or irreducible to anything else.
- I will call this the "principle of irreducibility", but it is a prince that does not govern since that would be a self-contradiction (2.6.1).

1.1.2 There are only trials of strength, of weakness. Or more simply, there are only trials. This is my point of departure: a verb, "to try."

1.1.3 It is because nothing is, by itself, reducible or irreducible to anything else that there are only trials (of strength, of weakness). What is neither reducible nor irreducible has to be tested, counted, and measured. There is no other way.

1.1.4 Everything may be made to be the measure of everything else.

1.1.5 Whatever resists trials is real.
- The verb "resist" is not a privileged word. I use it to represent the

whole collection of verbs and adjectives, tools and instruments, which together define the ways of being real. We could equally well say "curdle", "fold", "obscure", "sharpen", "slide." There are dozens of alternatives.

1.1.5.1 The real is not one thing among others but rather gradients of resistance.

1.1.5.2 There is no difference between the "real" and the "unreal", the "real" and the "possible", the "real" and the "imaginary." Rather, there are all the differences experienced between those that resist for long and those that do not, those that resist courageously and those that do not, those that know how to ally or isolate themselves and those that do not.

1.1.5.3 No force can, as it is often put, "know reality," other than through the difference it creates in resisting others.
- In the old days it would have been said that force and knowledge are coextensive, or, as in the fable, that "the strongest reason always yields to reasons of the strongest."

1.1.5.4 Nothing is known—only realized.

1.1.6 A shape is the front line of a trial of strength that de-forms, trans-forms, in-forms or per-forms it. Of course, once a form is stable, it no longer appears to be a trial of strength.

1.1.7 What is a force? Who is it? What is it capable of? Is it a subject, text, object, energy, or thing? How many forces are there? Who is strong and who is weak? Is this a battle? Is this a game? Is this a market? All these questions are defined and deformed only in further trials.
- In place of "force" we may talk of "weaknesses", "entelechies", "monads", or more simply "actants."

1.1.8 No actant is so weak that it cannot enlist another. Then the two join together and become one for a third actant, which they can therefore move more easily. An eddy is formed, and it grows by becoming many others.

■ Is an actant essence or relation? We cannot tell without a trial (1.1.5.2). To stop themselves being swept away, essences may relate themselves to many allies, and relations to many essences.

1.1.9   An actant can gain strength only by associating with others. Thus it speaks in their names. Why don't the others speak for themselves? Because they are mute; because they have been silenced; because they became inaudible by talking at the same time. Thus, someone interprets them and speaks in their place. But who? Who speaks? Them or it? *Traditore—traduttore.* One equals several. It cannot be determined. If the fidelity of the actant is questioned, it can demonstrate that it just repeats what the others wanted it to say. It offers an exegesis on the state of forces, which cannot be contested even provisionally without another alliance.

1.1.10   Act as you wish, so long as this cannot be easily undone. As a result of the actants' work, certain things do not return to their original state. A shape is set, like a crease. It can be called a trap, a ratchet, an irreversibility, a Maxwell's demon, a reification. The exact word does not matter so long as it designates an asymmetry. Then you cannot act as you wish. There are winners and losers, there are directions, and some are made stronger than others.

1.1.11   Everything is still at stake. However, since many players are trying to make the game irreversible and doing everything they can to ensure that everything is not equally possible, the game is over.
■ Homage to the *Masters of Go* (Kawabata: 1972).

1.1.12   To create an asymmetry, an actant need only lean on a force slightly more durable than itself. Even if this difference is tiny, it is enough to create a gradient of resistance that makes them both more real for another actant (1.1.5).

1.1.13   We cannot say that an actant follows rules, laws, or structures, but neither can we say that it acts without these. By learning from what the other actants do, it gradually elaborates rules, laws, and structures. Then it seeks to make the others play by these rules which it claims to have learned, observed, or received. If it wins, then it verifies them and has thereby applied them.
■ Is any given order a convention, a social construction, a law of

nature, or a structure of the human mind? We cannot say. But in love as in war all is fair in the attempt to attach the rules to something more durable than the moment that inspired them.

1.1.14 Nothing is by itself ordered or disordered, unique or multiple, homogeneous or heterogeneous, fluid or inert, human or inhuman, useful or useless. Never by itself, but always by others.

- Spinoza said it long ago: so far as *shapes* are concerned, let us not be anthropo*morphic*. Each weakness distributes a complete range of roles. Depending on what it expects from the others, it distinguishes the stable and the ordered from the shapeless and the moving. But since the others all distribute roles as well, a beautiful tangle ensues. Still, it is comprehensible why entelechies may mistake those they broke down, dismembered, or seduced for shapeless matter.

1.1.14.1 Order is extracted not from disorder but from orders.

- We always make the same mistake. We distinguish between the barbarous and the civilized, the constructed and the dissolved, the ordered and the disordered. We are always lamenting decadence and the dissolution of morals. Bad luck! Attila speaks Greek and Latin; punks dress with the same care as Coco Chanel; plague bacteria have strategies as subtle as those of IBM; the Azande falsify their beliefs with the gusto of a Popper. No matter how far we go, there are always forms; within each fish there are ponds full of fish. Some believe themselves to be the molds while others are the raw material, but this is a form of elitism. In order to enroll a force we must conspire with it. It can never be punched out like sheet metal or poured as in a cast.

1.1.15 "Everything is necessary" and "everything is contingent" mean the same thing—that is nothing. The words "necessary" or "contingent" gain meaning only when they are used in the heat of the moment to describe gradients of resistance—that is, reality.

- The length of Cleopatra's nose is neither significant nor insignificant. Circumstances determine, for a time, the relative importance of whatever it is that makes them up. Chance and necessity cannot be allocated their roles in advance.

1.1.16 What is the same and what is different? What is with whom? What is opposed or allied or intimate? What continues, stops, abandons, hastens, or attaches itself? These are common questions, yes, *common* to all trials whether we fawn, taste, unravel, plait, join, erase, or address.

1.2.1   Nothing is, by itself, the same as or different from anything else. That is, there are no equivalents, only translations.

In other words, everything happens only once, and at one place.

If there are identities between actants, this is because they have been constructed at great expense. If there are equivalences, this is because they have been built out of bits and pieces with much toil and sweat, and because they are maintained by force. If there are exchanges, these are always unequal and cost a fortune both to establish and to maintain.

- I call this the "principle of relativity." Just as it is not possible for one observer to communicate with another more quickly than the speed of light, the best that can be done between actants is to translate the one into the other. There is nothing between incommensurable and irreducible forces: no ether, no instantaneousness. It is true that this principle of relativity aims to reestablish the inequivalence of actants, whereas the other principle was designed to restore the equivalence of all observers. In both, however, we have to get used to breathing in the absence of the ether. The stuff of which I speak is rare, dispersed, and mostly empty. Gatherings, saturations, and plenitudes are uncommon and dispersed, like large towns on the map of a country.

Interlude 1: In a Pseudoautobiographical Style to Explain the Aims of the Author

I taught at Gray in the French provinces for a year. At the end of the winter of 1972, on the road from Dijon to Gray, I was forced to stop, brought to my senses after an overdose of reductionism. A Christian loves a God who is capable of reducing the world to himself because he created it. A Catholic confines the world to the history of the Roman salvation. An astronomer looks for the origins of the universe by deducing its evolution from the Big Bang. A mathematician seeks axioms that imply all the others as corrolaries and consequences. A philosopher hopes to find the radical foundation which makes all the rest epiphenomenal. A Hegelian wishes to squeeze from events something already inherent in them. A Kantian reduces things to grains of dust and then reassembles them with synthetic a-priori judgments that are as fecund as a mule. A French engineer attributes potency to calculations, though these come from the practice of an old-boy network. An administrator never tires of looking for officers, followers, and subjects. An intellectual strives to make the "simple" practices and opinions of the vulgar explicit and conscious. A son of the bourgeoisie sees the simple stages of an abstract cycle of wealth in the vine growers, cellarmen, and bookkeepers. A Westerner never tires of shrinking the evolution of species and empires to Cleopatra's nose, Achilles' heel, and Nelson's blind eye. A writer tries to recreate daily life and imitate nature. A painter is obsessed by the desire to render feelings into colors. A follower of Roland Barthes tries to turn everything not only into texts but into signifiers alone. A man likes to use the term "he" in place of humanity. A militant hopes that revolution will wrench the future from the past. A philosopher sharpens the "epistemological break" to guillotine those who have not yet "found the sure path of a science." An alchemist would like to hold the philosopher's stone in his hand.

To put everything into nothing, to deduce everything from almost nothing, to put into hierarchies, to command and to obey, to be profound or superior, to collect objects and force them into a tiny space, whether they be subjects, signifiers, classes, Gods, axioms—to have for companions, like those of my caste, only the Dragon of Nothingness and the Dragon of Totality. Tired and weary, suddenly I felt that everything was still left out. Christian, philosopher, intellectual, bourgeois, male, provincial, and French, I decided to make space and allow the things which I spoke about the room that they needed to "stand at arm's length." I knew nothing, then, of what I am writing now but simply repeated to myself: "Nothing can be reduced to anything else, nothing can be deduced from anything else, everything may be allied to everything else." This was like an exorcism that defeated demons one by one. It was a wintry sky, and a very blue. I no longer needed to prop it up with a cosmology, put it in a picture, render it in writing, measure it in a meteorological article, or place it on a Titan to prevent it falling on my head. I added it to other skies in other places and reduced none of them to it, and it to none of them. It "stood at arm's length," fled, and established itself where it alone defined its place and its aims, neither knowable nor unknowable. It and me, them and us, we mutually defined ourselves. And for the first time in my life I saw things unreduced and set free.

1.2.2 Entelechies agree about nothing and can agree on everything, for nothing is, in and of itself, either commensurable or incommensurable. Whatever the agreement, there is always something upon which disagreement may feed. Whatever the distance, there is always something upon which an understanding may be built. To put it another way, everything is negotiable.

- "Negotiation" is not a bad word so long as it is understood that everything is negotiable, not just the shape of the table or the names of the delegates. Decisions also have to be made on what the negotiation is all about, when it can be said to have started or finished, what language will be spoken, and how whether we have been understood or not will be determined. Was it a battle, a ceremony, a discussion, or a game? This is also a matter of dispute, a dispute that continues until all the entelechies are defined and have themselves defined the others. It is to display these negotiations that I need a Field of the Cloth of Gold.

1.2.3   How many actants are there? This cannot be determined until they have been measured against each other.

- I have not yet said how many we were: 50 million Frenchmen, a single ecosystem, 20 billion neurons, three or four types of character, a single "me, I, me, I." We cannot count the number of forces, decide that there is a unique substance, two social classes, three graces, four elements, seven deadly sins, or twelve apostles. We cannot add up a total. In this peculiar arithmetic no one ever subtracts. We add as many subtotals as there are accountants.

1.2.3.1   There are neither wholes nor parts. Neither is there harmony, composition, integration, or system (1.1.14). How something holds together is determined on the field of battle, for no one agrees who should obey and who command, who should be a part and who the whole.

- There is no preestablished harmony, Leibniz notwithstanding, harmony is *post*established locally through tinkering.

1.2.4   We do not know where an actant is to be found. The definition of its location is a primordial struggle, during which many get lost. We can only say that some locate and others are located.

1.2.4.1   Though places are distant, irreducible, and unsummable, they are nevertheless constantly brought together, united, added up, aligned, and subjected to ways and means. If it were not for these ways and means, no place would lead to any other.

1.2.5   Forces that ally themselves in the course of a trial are said to be durable. Each entelechy generates times for others by allying with or betraying them. "Time" arises at the end of this game, a game in which most lose what they have staked.

■ Is this moment before or is it after? Is it overtaken, prophetic, obsolete, decadent, contemporary, provisional, or eternal? This cannot be determined in advance. It has to be negotiated.

1.2.5.1 Time is the distant consequences of actors as they each seek to create a fait accompli on their own behalf that cannot be reversed (1.1.10). In this way time passes.

1.2.5.2 Time does not pass. Times are what are at stake between forces. Of course, one force may overtake the others, but this can only be local and temporary because permanence costs too much and requires too many allies.

1.2.5.3 It is often said in France that "there are" revolutions, but these are only actors which take their capacity to make time and history from other actors and thereby pass the others by and make them passé. Of course, the vanquished sometimes obtain their revenge and thus upset the order of times once more.

■ Who, then, is the most modern—the Shah; Khomeini, the Muslim from another age; or Bani-Sadr, the President, who has sought refuge in Paris? No one knows, and this is why they struggle so much to make their time.

1.2.5.4 The freest of all democracies reigns between instants. No instant can crown, cripple, justify, replace, or limit any other. There is no last moment to condemn all those that came before.

■ Times are irreducible, and this is why "death" has always been vanquished. The end does not justify the means. Neither does death condemn life.

1.2.6 Space and time do not frame entelechies. They only become frameworks of description for those actants that have submitted, locally and provisionally, to the hegemony of another.

■ There is therefore a time of times and a space of spaces, and so on until everything has been negotiated. Homage to Péguy's *Clio* (1914).

1.2.7 Each entelechy defines: what lies inside it and what outside, which other actors it will believe when it decides what belongs to it and what does not, and which kinds of trials it will use to decide whether or not to believe these referees.

- Leibniz was right to say that monads have neither doors nor windows, for they never come out of themselves. However, they are sieves, for they endlessly negotiate about their frontiers, about who the negotiators will be, and about what they ought to do. As a result they end up like chimeras, unable to determine which is the door and which the window, which is stage left and which stage right.

1.2.7.1   There is no external referent. Referents are always internal to the forces that use them as touchstones.

1.2.7.2   The principle of reality is other people.

- The interpretation of the real cannot be distinguished from the real itself because the real are gradients of resistance (1.1.5). An actant therefore never stops negotiating the number, the gradient, and the nature of these differences; the number, the authority, and the weight of those who negotiate; the number, the quality, and the reliability of the touchstones that they will use to judge the credibility of the referees.

1.2.8   Every entelechy makes a whole world for itself. It locates itself and all the others; it decides which forces it is composed of; it generates its own time; it designates those who will be its principle of reality. It translates all the other forces on its own behalf, and it seeks to make them accept the version of itself that it would like them to translate.

- Nietzsche called this "evaluation," and Leibniz "expression."

1.2.9   Is it a force of which we speak? Is it a force that speaks? Is it an actor made to speak by another? Is it an interpretation or the object itself? Is it a text or a world? We cannot tell, because this is what we struggle about, the building of a whole word.

- What those who use hermeneutics, exegesis, or semiotics say of texts can be said of all weaknesses. For a long time it has been agreed that the relationship between one text and another is always a matter for interpretation. Why not accept that this is also true between so-called texts and so-called objects, and even between so-called objects themselves?

1.2.10 Nothing escapes the primordial trials. Before negotiation we have no idea what kind of trials there will be—whether they can be thought of as conflict, game, love, history, economy, or life. Neither do we know whether they are primordial or secondary before we enter the arena. Finally, we cannot tell until the end whether they have been negotiated or were received at birth, etched into the skin itself.

1.2.11 We must not believe in advance that we know whether we are talking about subjects or objects, men or gods, animals, atoms, or texts. I have not yet said, for this is precisely what is at stake between forces: who speaks, and of what?
- We should not hurry to divide "nature" from "culture." Scallops also find that nature is a harsh taskmaster—hostile, nourishing, profligate—because fish, fishermen, and the rocks to which they attach themselves have ends that differ from those of scallops.

1.2.12 Nothing is, by itself, either knowable or unknowable, sayable or unsayable, near or far. Everything is translated. What could be simpler?

1.2.13 If everything we have to write about is to be debated and translated, then we need, as Descartes said, a provisional moral. When we speak of trials of strength, we must avoid using any terms that fix the relationship to the advantage of one side or the other. If this is not possible, we should at least try to write a text that does not take time and space but provides it instead.

1.3.1 All entelechies may measure and be the measure of all other entelechies (1.1.14). Nevertheless, certain forces constantly try to measure rather than be measured and to translate rather than be translated. They wish to act rather than be acted upon. They wish to be stronger than the others.
- I have said "certain" rather than "all" as in Nietzsche's bellicose myth. Most actants are too far apart or too indifferent to rise to the challenge, too undisciplined or devious to follow for long those that speak in their name, and too happy and proud to take command of others. In this work I speak only of those weaknesses that want to increase their strength. The irreducible others have need of poets rather than philosophers.

1.3.2   Given that actants are incommensurable and that each makes a world as large and complete as any other, how does it happen that one becomes more than another? By claiming to be several, by associating (1.1.9).

1.3.3   Since nothing is, in and of itself, either equivalent, or not equivalent (1.2.1), two forces cannot associate without misunderstanding.

- Entente, arrangement, compromise, negotiation, scheme, combination, compact—all these terms can be used. Those who find them derogatory and believe that they conflict with more perfect forms of association fail to understand that it is never possible to do better, both because there is no equivalence (2.2.1) and because nothing is, by itself, either reducible or irreducible to anything else (1.1.1).

1.3.4   Although all entelechies are "equally" active, they may appear to be in two states: dominating or dominated, acting on or acted upon. For an entelechy to be called passive, it need only fail to answer back.

- I am not saying that there are active forces and ones that are passive, but only that one force may act as if another were passive and obedient (1.1.14). For the passive force, of course, the point of view is entirely different. There are a thousand reasons for feigning obedience, ten thousand for wishing to be dominated, and a hundred thousand for remaining silent—reasons that are never suspected by those who believe they are served.

1.3.5   Since an actant can become greater than another only by being one of several, and since this association is always a misunderstanding, the one who defines the nature of the association without being contradicted takes control.

- Where two forces proclaim themselves to be united, only one speaks; where two forces makes an exchange they deem to be equal, one always determines who defines the thing exchanged, how equality is measured, and when the exchange has taken place.

1.3.6   Since nothing is equivalent, to be strong is to make equivalent what was not. In this way several act as one.

- "Anything does not go." Discourses and associations are not equivalent, because allies and arguments are enlisted precisely so that one association will be stronger than another. If all discourse appears to be equivalent, if there seem to be "language games" and nothing more, then someone has been unconvincing. This is the weak point of the relativists. They talk only about forces that are incapable of allying themselves with others in order to convince and win. By repeating "anything goes," they miss the work that generates inequivalence and asymmetry (1.1.11).

1.3.7 Since nothing is commensurable or incommensurable (1.1.4), the more active is the one that is able to define the mechanisms of measurement.

- There are *acts* of differentiation and identification, not differences and identities (1.1.16). The words "same" and "other" are the consequences of trials of strength, defeats and victories. They cannot themselves describe these links.

Interlude II: Showing What a Relief It Is to Stop Reducing Things

Sometimes when the sun shines on the roughened concrete of the Salk Institute, we stop hurrying about and using up time. We sit on our doorstep and let each branch of the tree of times unfold as far as it can. "Nothing is by itself either reducible or irreducible to anything else," we say of all those who reduce, destroy, replace, deduce, permutate, explain, cause, redeem, restore, involve, determine, exchange, and buy. The tree of times, the trees of times, the forest of trees of times. Nothing is changed, yet the position of each force, each entelechy, each actor changes so completely that we breathe an air that we did not know we were missing before.

At these moments it is not the being as being that reveals itself. This business of being as *being* has become quite incongruous now that each entelechy has all the differences it needs to make a whole world for itself. The tide has changed. Before there were only things that had been reduced and things that did the reducing, with a residual being who rattled around in our heads like a pea in a pod. Does this mean there is fusion, ataraxia, or lack of differentiation? No, of course not! All the differences are there. Not a single one is missing. And all the attempts to reduce, produce, simplify, hierarchize, totalize, or destroy them are likewise there, like so many differences which add themselves to those that they wished to suppress.

Nothing pardons, makes amends for, atones, balances, succeeds, subsumes, concludes, summarizes, or submits to itself. And yet we should indeed speak about a state of grace. Everything is light, for nothing has the power to bring about the dizzy fall of anything else. Yes, freedom to go, freedom to do, freedom to pass, freedom to let go. The seagull, far from its name, far from its species, in its own world of air, sea, and favored fish; the fish far from its shoals, far from the gull and its beak, innocent in the icy water; the water that gathers together and shapes itself, mixed by the winds, knotted by the currents, heaving and breaking itself onto the beach; the oceanographer turned frogman who dives into the La Jolla submarine canyon; the managing director who produces *Jaws II* after *Jaws* and sells fear of the deep and the shadow of sharks—all are innocent. Innocent? No. Neither innocent nor guilty. Marked, inscribed, unpardonable. When the tree of times is left to grow, the act and its consequences are separated, and each becomes the means and the end of the other. It is thus impossible to atone for a means with an end, for a life of crime with a prayer, for a man with his children, for a managing director with his bank account. No equivalences, no market. We can neither die nor conquer death. There is room for the one who has lived, for the day of her death, for the bullet of the killer, for the inquiry that leads to no conclusion, for the memory of those who speak of the dead friend. Nothing sums up those places, nothing explains them, nothing justifies them. Innocent? No, since we have gone beyond the distinction between the innocent and the guilty made by the erection of the scaffold. Incomprehensible? No, since we are beyond operations that establish, day after day, what we understand and what we do not know. The bird, far from its name, flies from the name that I give it, but continues to fly in treatises on zoology and the poems of St. John Perse. The gull is in its sky, irreducible to ours, but the language of the taxonomist is in the books, itself irreducible to any gull ever dreamed of, living or dead.

1.4.1   Certain actants test their strength against others, declare them to be passive, and make an alliance with them that they themselves define. By imposing equivalences which they direct, they spread themselves step by step from passive actor to passive actor.

- We too often tend to start with "exchanges," "equalities," and the "transfer" of equivalents. But we never talk about the preliminary work in which these equivalents are forged. It is as if we spoke of road networks but never of civil engineering. However, there is as much of a difference between equivalent and making equivalent as between driving an automobile and building a freeway.

1.4.2   When one weakness enlists others, it forms a network so long as it is able to retain the privilege of defining their association.

- In a network certain very distant points can find themselves connected, whilst others that were neighbors are far removed from one another. Though each actor is local, it can move from place to place, at least as long as it is able to negotiate equivalences that make one place the same as another. A network can thus be "quite general" without ever having to pass through a "universal." However rarefied and convoluted a network may be, it nevertheless remains local and circumscribed, thin and fragile, interspersed by space. We should imagine filamentlike entelechies, spun out and interwoven with one another (1.2.7), which are incapable of harmony because each one defines the size, the tempo, and the orchestration of this harmony.

1.4.3 Between one network and another, as between one force and another (1.2.7), nothing is by itself either commensurable or incommensurable. Thus we never emerge from a network no matter how far it extends.

- It is for this reason that one can be Commandant at Auschwitz, an olive tree at Corfu, a plumber in Rochester, a seagull in the Isles of Scilly, a physicist at Stanford, gneiss in the Minas Gerais, a whale in Adelie Land, one of Koch's baccili at Damiette, and so on. Each network makes a whole world for itself, a world whose inside is nothing but the internal secretions of those who elaborate it. Nothing can enter the galleries of such a network without being turned outside in. If we thought that termites were better philosophers than Leibniz, we could compare a network to a termites' nest—so long as we understood that there is no sun outside to darken its galleries by contrast. It will never be possible to see more clearly, it will never be possible to get further "outside" than a termite, and the most widely accepted equivalence might appear, under trial, no stronger than a wall of clay.

1.4.4 A force establishes a pathway by making other forces passive. It can then move to places that do not belong to it and treat them as if they were its own.

- I am willing to talk about "logic" (2.0.0), but only if it is seen as a branch of public works or civil engineering. To speak in this way is more accurate than to talk, like Ulrich, of a General Secretariat for Precision and the Spirit (Musil, ch. 116).

1.4.5 Entelechies wishing to be stronger can be said to create *lines of force*. They keep others in line. They make them more predictable.

- The term "line of force" is even vaguer than "network," "way," "gallery," or "logic," but this is fine. The reader should not yet be able to decide whether I am speaking of social beings, printed circuits, reasons, machines, theaters, or habits. This vagueness is exactly the effect I am seeking, for perhaps we will never come across objects classified in this way again.

1.4.6 As soon as one actant manages to persuade others to fall into line, it thereby increases its strength and becomes stronger than those it aligned and convinced (1.5.1). This gain can be measured in a number of ways. It can be said that A is *connected* to others. Although in principle every connection is equally possible, it now becomes easier to link B to A than to C. A can also be said to *command* others. Although in principle these others lend their strength to A, they allow themselves to be controlled by it. A can also be said to *translate* the wishes of others. Although the others might wish to say something else, they agree that what A says is what they wanted to say but were not able to put into words. A's strength can also be measured by saying that it can *buy* others. Although in principle the others are not worth the same amount (1.2.1), E or F agree to be equivalent to what A is ready to pay. Finally, it can be said that A *explains* others. Although the others cannot reduce themselves to A, they agree to be its consequences, predicates, or applications (2.0.0).

In the final reckoning the work of making value and making equivalent means that A is stronger than others despite their incommensurability. It translates, explains, understands, controls, buys, decides, convinces, and makes them work.

- Sometimes this accumulation of equivalents or tokens is called "capital," but capital was not the initial step. First it was necessary to create equivalences (1.3.7), bend forces, and hold them in place for long enough to be scaled and measured. Only then was it possible to calculate a profit (1.3.5). The marketplace is only a consequence of the establishment of networks; it does not explain their formation.

1.4.6.1 An absolute force is one that would be capable of explaining everything, translating everything, producing everything, buying and redeeming everything, and causing everything to act. As a universal equivalent, capable of substituting itself for everything, and a universal providence, capable of giving life to everything, it would be the prime mover and first principle from which all the rest could be generated.

- Some people talk of "God" when they think of the force that is capable of redeeming the world by His Son, of explaining the origin and the creation, of translating into His word what every creature, animate and inanimate, wishes at the bottom of its heart, of shepherding us through the detours of Providence to that which we all desire. Because nothing is by itself either reducible or irreducible (1.1.1), this absolute force is also the absolutely pure expression of nothingness. Because of its very purity it has always fascinated mystics, warlords, captains of industry, and scholars in search of first principles. "Oh", they all say to themselves, "grasp a single force (a town, a chalice, an axiom, a bank), and the rest shall be given unto us." To avoid the panic of reduction, we must always say: "What is left is all (Interlude I-II). The great Pan is dead."

1.4.6.2 An actor expands while it can convince others that it includes, protects, redeems, or understands them. It extends itself faster and further if it can secure actors who have already made themselves equivalent to many others.

- It has often been said that "capitalism" was a radical novelty, an unheard-of rupture, a "deterritorialization" pushed to the ultimate extreme. As always, the Difference is mystification. Like God, capitalism does not exist. There are no equivalents (1.2.1); these have to be made, and they are expensive, do not lead far, and do not last for very long. We can, at best, make extended networks (1.4.2). Capitalism is still marginal even today. Soon people will realize that it is universal only in the imagination of its enemies and advocates (Interlude VI). Just as Roman Catholics believe in the universality of their religion even though it only flows in Roman channels, the enemies and supporters of capitalism believe in what is perhaps the purest of mystical dreams: that an absolute equivalence has been achieved. Even the United States, the country of true capitalism, cannot fully live up to its ideal. Despite the efforts of the trade unions and the employers' associations, forces swarm that cannot be made equivalent without work (3.0.0). My homage to Fernand Braudel (1985), who does not hide this fact and shows how long-distance control may be achieved through tenuous networks.

1.5.1 A force cannot be *given* those forces that it arrays and convinces. By definition it can only *borrow their support* (1.3.4). Nevertheless, it will claim what does not belong to it and will add their forces to its own in a new form: in this way *potency is born.*

- When an entelechy contains other entelechies which it does not contain, we say that it contains them "potentially." The origin of potency lies in this confusion: *it is no longer possible to distinguish an actor from the allies which make it strong.* From this point on we begin to say that an axiom implies its demonstration "in potentia"; we begin to say of a prince that he is powerful, that the being-in-itself contains the being for itself, though only "potentially." With potency injustice also begins, because apart from a happy few—princes, principles, origins, bankers, and directors—other entelechies, that is, all the remainder, become details, consequences, applications, followers, servants, agents—in short, the rank and file. Monads are born free (1.2.8), and everywhere they remain in chains.

1.5.1.1   Talk of possibilities is the illusion of actors that move while forgetting the cost of transport.
- Producing possibilities is as costly, local, and down to earth as making special steels or lasers. Possibilities are bought and sold like everything else. They are not different by nature. They are not, for example, "unreal." There is no such thing as a free possibility. The files of consultants are expensive—ask those who went bankrupt because they produced too many possibilities but did not sell enough.

1.5.2   If an actor contains many other in potentia, it is impressive because, even when alone, it is a crowd. That is why it is able to enroll other actors and borrow their support more easily.
- Although it starts out as a bluff by claiming to own what has only been borrowed, it becomes real. Since the real is what resists (1.1.5), who is able to resist an entelechy turned crowd? Powers, thrones, and dominations spread rank after rank, though they have neither grown nor moved and are as weak as those who allow them to act.

1.5.3   Power is never *possessed.* We either have it "in potentia," but

then we do not *have* it; or we have it "in actu", but then our allies
are the ones that go into action.

- The philosophers and sociologists of power flatter the masters they
  claim to criticize. They explain the masters' actions in terms of the
  might of power, though this power is efficacious only as a result
  of complicities, connivances, compromises, and mixtures (3.4.0)
  which are not explained by power. The notion of "power" is the
  dormitive virtue of the poppy which induces somnolence in the
  critics at just the moment when powerless princes ally themselves
  with others who are equally weak in order to become strong.

1.5.4 Though they can neither count nor sum the others up, fewer
and fewer forces with nothing of their own attribute the potency of
all other powers to themselves. This is the reductio ad absurdum of
the whole to nothing. Princes who are almost nothing act as if the
rest, that is, everything, were no longer anything.

Chapter 2
Sociologics

2.1.1 All reasoning is of the same form: one sentence follows another. Then a third asserts that these are identical even though they do not resemble one another. Thenceforth the second is used in place of the first, and a fifth affirms that the second and the fourth are identical, even though . . . and so on, until one sentence is *displaced* while pretending not to have moved, and *translated* while pretending to have stayed faithful.

2.1.2 There has never been such a thing as deduction. One sentence *follows* another, and then a third affirms that the second was implicitly or potentially already *in* the first (1.5.1).
- Those who talk of synthetic a-priori judgments deride the faithful who bathe at Lourdes. However, it is no less bizarre to claim that a conclusion lies *in* its premises than to believe that there is holiness *in* the water.

2.1.3 When many different sentences have been made equivalent, they are all folded back into the first, of which it is said that this "implies them all." This single phrase is then bandied about, and it is claimed that all the others may be extracted from it "by pure deduction."

2.1.3.1 Those who reason in front of others and claim to extract one phrase from *another* are at best jugglers and at worst cheats. For years they have been practicing their tricks using rabbits and hats borrowed from onlookers.

2.1.3.2 Only teachers claim to be able to extract one sentence from another by means of "pure, formal deduction." They know in advance the conclusion of the argument that they claim to be unfolding. Organized arguments learned *slowly* and *in disorder* are unfolded by them at high speed, one after another, concealing what went on backstage behind the blackboard, the tumultuous history that led this proposition to be linked to that one. They offer that which contains in potentia all the consequences for the worship of their pupils, who fervently believe that they have deduced one thing from another.

- Without schooling, no one would have faith in this religion of deduction. We might as well say that the propositions of Spinoza's *Ethics* are "all in" the first proposition, or that the dessert is contained in the entrée. But schoolboys have always been fascinated by the absolute cribs offered by Laplace's principle: to hold all knowledge in the palm of our hand, having extracted it from the heel of our shoe.

2.1.4 Arguments form a system or structure only if we forget to test them. What? If I were to attack *one* element, would *all* the others then come crowding round me without a moment's hesitation? This is so unlikely! Every collection of actants include the lazy, the cowardly, the double agents, the dreamers, the indifferent, and the dissidents. Yes, I grant you that the fear of seeing A, B, or E coming to the rescue can so impress people that they give up. But if they hold on, the odds are that B will be dissociated, because C comes too slowly, E is depressed, F is a traitor, and G was unable to help because it was trying to stop F's betrayal.

- As is well known, an alliance between the logicians and the army led General Stumm to put the solidity of structures to the test in the library at Vienna (Musil, ch. 85). He was very disappointed.

In Paris we still believe in structures because we take care not to test their loyalty.

2.1.5   Commentary is never faithful. Either there is repetition, which is not commentary, or there is commentary, which is said *differently*. In other words there is translation and betrayal. Despite this, exegetes never tire of imputing glosses to the text. The text is puffed up with all the glosses that it has to contain "in potentia" in order to justify all these readings.

- Texts are never faithful to one another, but always at some distance.

2.1.6   We say "whoever controls the cause, controls the effect," as if the effect were potentially contained within the cause. However, no *word* can cause another. Words *follow* one another in a story. It is only later in the story that one character is made the "cause" and another the "consequence." The only effect to consider is the effect upon the public of this or that alliance of words: "No, he's exaggerating," or "it's well written," or again, "very illuminating," "very convincing," "how full of himself," or "what a bore."

2.1.7   There are no theories. There are texts to which, like lazy potentates, we respectfully attribute things that they have not done, inferred, foreseen, or caused. Theories are never found alone, just as in open country there are no clover leaf intersections without freeways to connect and redirect.

2.1.7.1   In theory, theories exist. In practice, they do not.

- No one has ever deduced all of geometry from the axioms and postulates of Euclid. But "in theory," they say, "anyone can anywhere" derive "the whole of" geometry "at any time" from the axioms of Euclid "alone." In practice, this has *never happened* to *anyone*. But no one has ever needed to draw this conclusion, because "in theory" the opposite remains possible. And sorcerers are scorned because they are said to be incapable of accepting facts even when facts have contradicted them every day for centuries!

2.1.7.2   There is no *meta*language, only infralanguages. In other words there are only languages. We can no more reduce one language to another than build the tower of Babel.

- Those who talk of metalanguage must mean, I think, the pidgin of the masters which is too impoverished even to translate what is said in the kitchen.

2.1.7.3 Daily practice needs no theorist to reveal its "underlying structure." "Consciousness" does not underlie practice but is something else somewhere else in another network. Practice lacks nothing.
- *Where* are the unconscious structures of primitive myths? In Africa? In Brazil? No! They are among the filing cards of Lévi-Strauss's office. If they extend beyond the Collège de France at the rue des Ecoles, it is through his books and disciples. If they are found in Bahia or Libreville, it is because they are taught there.

2.1.8 So far as form is concerned (2.1.1), all arguments are equally good. All that we need is a series of sentences, and then we say that some are the same and others different (2.1.2). The sentences are then woven into plaits, tresses, garlands, wreathes, and webs. This can *always* be done, can't it? As a result, *certain* moves become easier and others more difficult.
- No one can classify arguments in terms of their *formal* qualities. If you insist, we may rank them in terms of their *material* qualities.

2.1.8.1 Nothing is by itself either logical or illogical. A path always goes somewhere. All we need to know is where it goes and what kind of traffic it has to carry. Who would be so foolish as to call freeways "logical," roads "illogical," and donkey tracks "absurd"?

2.1.8.2 No set of sentences is by itself either consistent or inconsistent (1.1.14); all that we need to know is who tests it with which allies and for how long. Consistency is felt (1.1.2); it is not a diploma, a medal, or a trademark.

2.1.8.3 The thread of *argument* is never straight. Those who talk of "logic" have never looked how something is spun, plaited, ranked, woven, or deduced. A butterfly flies in a straighter line than a mind that reasons. (Sometimes, of course, woven patterns may represent a straight line which is pretty to look at.)

2.1.8.4 "Reason" is applied to the work (2.5.4) of allocating agreement and disagreement between words. It is a matter of taste and feeling, know-how and connoisseurship, class and status. We insult, frown, pout, clench our fists, enthuse, spit, sigh, and dream. Who reasons?

- An anthropologist of body language could sketch the thinking of a Cambridge don or a Wall Street banker.

2.1.9 Since the *amount* of identities and differences that we have to *share* remains constant (2.1.8), it is not within our power to be illogical or irrational (2.1.8.1). Still, there are many ways to allocate "in consequences", "because ofs", "in contradiction withs," and "neverthelesses." No one is more attentive to "non sequiturs" than logicians, wizards, or stage managers. When effects are to be contrived, we have to choose what will follow what with great care. We have to decide when the name of the traitor or the axiom will be made known and prepare for the entry that will most impress the audience. We have to determine units of time and place, causes, and principles. We have to choose to write "more geometrico" or "more populo" as we tastefully select the theorems and asides. In brief, conviction depends on the genre we choose.

- We are forgetting that there are just as many skeptics, ratiocinators, Popperians, and rationalists among the Azande as there are among the Copernicuses and Szilards. Since the amount of agreements and disagreements is constant, we cannot *cleanly* separate mythical fictions from scientific accounts. This can be done only in a dirty way, and then it is real butchery. A painter who chooses only shades of gray is no less a painter than one who uses dazzling colors. There are proofs as rigorous as winter and there are springlike proofs, but they are all still proofs.

2.1.10 Since nothing is inherent in anything else, the dialectic is a fairy tale. Contradictions are negotiated like the rest. They are built, not given.

2.1.11 If magic is the body of practice which gives certain words the potency to act upon "things," then the world of logic, deduction, and theory must be called "magical": but it is *our* magic.

- Just as the Greeks called the fine languages of the Parthians, the Abyssinians, or the Sarmatans "barbaric," so we call the perfect arguments (2.1.8) of those who believe in other powers of deduction "illogical."

2.2.1 To say something is to say it in other words. In other words, it is to translate.

- A word is put in the place of another which it does not resemble. A third word says that they are the same (2.1.1). A is not A, but B and C. Rome is no longer in Rome, but in Crete and among the Saxons. This is called "predication." *That is to say,* we cannot speak properly, moving from the *same* to the *same*, but only roughly, moving from the same to the *other*.

2.2.2 Since nothing is reducible or irreducible to anything else (1.1.1) and there are no equivalences (1.2.1), every pair of words may be said to be identical or to have nothing in common. Thus, there are no clear ways of distinguishing literal from *figurative* meanings (Hesse: 1974). Every group of words may be dirty, exact, metaphorical, allegorical, technical, correct, or far-fetched.

2.2.3 Nothing is by itself either "sayable" or "unsayable." Everything is translated (1.2.12). Since one word always lends its sense to another from which it nevertheless differs, it is no more in our power to speak rightly or wrongly than to stop the little mill of the fairy tale from grinding out salt.

2.2.4 Either the same thing is said and nothing is said, or something is said but it is something else. A choice must be made. It all depends on the distance that we are prepared to cover and the forces that we are prepared to coax as we try to make words that are infinitely distant equivalent.

2.2.5 We may be understood, that is surrounded, diverted, betrayed, displaced, transmitted, but we are never understood *well*. If a message is transported, then it is transformed. We never get a message that is simply spread.

2.3.1 We never begin to talk in words that freely associate, but rather in our mother tongue (2.2.2).

- Others have already played with the words when we start talking (1.1.10). Year after year, century after century, others have made certain associations of sounds, syllables, phrases, and arguments possible or impossible, correct or barbaric, proper or vulgar, false

or elegant, exact or nonsensical. Even though none of these group-
ings is as solid as claimed (2.1.4), if we wish to undo or remake
them, we become the object of blows, bad marks, caresses, gunfire,
or applause.

2.3.2   Though there is no proper or figurative meaning, it is possible
to appropriate a word, reduce its meanings and alliances, and link it
firmly to the service of another.

- Yet all the perfumes of Arabia will not sweeten this little metaphor
to make it figurative (2.2.2).

2.3.3   All associations of sounds, of words, and of sentences are
equivalent (2.1.8), but since they associate precisely so that they are
no longer equivalent to each other (1.3.6), in the end there are victors
and vanquished, strong and weak, sense and nonsense, and terms that
are literal and metaphorical.

2.3.4   Nothing is by itself either logical or illogical (1.2.8), but not
everything is equally convincing. There is only one rule: "Anything
goes"; say anything as long as those being talked to are convinced.
You say that to get from B to C, you have to pass through D and E?
If no others raise their voice to suggest other ways, then you have
been convincing. They go from B to C along the suggested path even
though no one wants to leave B for C and there are lots of different
routes that could be taken. Those you sought to convince have ac-
quiesced. For them, there is no more "Anything goes." That will have
to do, *for you will never do any better* (1.2.1).

2.3.5   We can say anything we please, and yet we cannot. As soon
as we have spoken and rallied words, other alliances become easier
or more difficult. Asymmetry grows with the flood of words; as meaning flows, slopes and plateaus are soon eroded. Alliances are formed
among words on the field of battle. We are believed, we are detested,
we are helped, we are betrayed. We are no longer in control of the
game. Some meanings are suggested, while others are taken away;
we are commented upon, deduced, understood, or ignored. That's it:
we can no longer say whatever we please.

2.4.1   How does one series of sentences become so much "stronger"
than another that the latter becomes "illogical," "absurd," "contra-

dictory," "fictitious," or "childish"? Like a force (1.3.2), an argument becomes stronger only by making use of whatever comes to hand. In this way we can force an actant to confess that this or that sentence is "contradictory" or "absurd," until no one can be found to make the argument illogical any longer.

- Rhetoric cannot account for the force of a sequence of sentences because, if it is called "rhetoric," then it is weak and has already lost (1.3.6). Logic cannot account for the force, since it attributes the victory that results from certain sentences to "formal" qualities common to all argument (2.1.0). Then again, semiotics remains inadequate because it persists in considering only texts or symbols instead of also dealing with "things in themselves."

2.4.2 Words are never found alone, nor surrounded only by other words; they would be inaudible.

- An actant can make an ally out of anything, since nothing is by itself either reducible or irreducible (1.1.1) and since there is no equivalence without the work of making equivalent (1.4.0). A word can thus enter into partnership with a meaning, a sequence of words, a statement, a neuron, a gesture, a wall, a machine, a face . . . anything, so long as differences in resistance allow one force to become more durable than another. Where is it written that a word may associate only with other words? Each time the solidity of a string of words is tested, we are measuring the *attachment* of walls, neurons, sentiments, gestures, hearts, minds, and wallets—that is, a heterogenous multitude of allies, mercenaries, friends, and courtesans. But we cannot stand this impurity and promiscuity.

2.4.3 We cannot distinguish between those moments when we have might and those when we are right.

- Trials of strength only sometimes take the form of a show of force (1.1.2); they also appear in many other guises. At one extreme actants operate so peacefully that they vanish into the background and become the flow of nature. Their action is so peaceful that no force seems to be exercised at all (1.1.6). At the other extreme there is bloodshed—total warfare without ritual, purpose, or preparation. Does this ever happen? Somewhere in between, I suppose, lies the great game of rhetoric, where the strength of a word may sway alliances and demonstrate something, where very, very rarely, *everything else being equal,* someone speaks and persuades. We always limit ourselves to talking about these three textbook cases; I want to talk about all the other cases as well.

2.4.4 Languages neither dominate nor are dominated, neither exist nor do not exist. They are entelechies like all others. They seek allies at their convenience and build a whole world from them with the same prohibitions and privileges as other actants.

- Only linguists could believe that words associate only with other words to make a linguistic structure. They forget the difficulty that they had in detaching words from their allies when they invented their structures. That words are forces like others with their own times and spaces, their "habitus" and their friendships, is surprising only to those who believe that "men" exist or dominate languages. Have you never fought with a word? Is not your tongue hardened by talking? Whatever resists is real (1.1.5). Who could believe that words have a clean history of their own?

2.4.5 It is not possible to distinguish for long between those actants that are going to play the role of "words" and those that will play the role of "things." If we talk only of languages and "language games," we have already lost, for we were absent when the changing roles and costumes were distributed.

- Recently there has been a tendency to privilege language. For a long time it was thought to be transparent, to be alone among actants in possessing neither density nor violence. Then doubts began to grow about its transparency. Hope was expressed that this transparency might be restored by cleaning language as we might clean a window. Language was so privileged that its critique became the only worthy task for generations of Kants and Wittgensteins. Then in the fifties it was realized that language was opaque, dense, and heavy. This discovery did not, however, mean that it lost its privileged status and was equated with the other forces that translate and are translated by it. On the contrary, the attempt was made to reduce all other forces to the signifier. The text was turned into "the object." This was "the swinging sixties," from Lévi-Strauss to Lacan by way of Barthes and Foucault. What a fuss! Everything that is said of the signifier is right, but it must also be said of every other kind of entelechy (1.2.9). There is nothing special about language that allows it to be distinguished from the rest for any length of time.

2.4.6 The consistency of an alliance is revealed by the number of actors that must be brought together to separate it (2.1.8.2). Therefore, we have to test it if we want to know what we are dealing with— if we want to know where the efficacy so often attributed to an isolated word, a solitary text, or a sign in the heavens actually *comes from*.

- They say, "You cannot go from B to D without passing through C or E." "If you are uncertain about C, then you are also in doubt about B and D." "If you are at B, you must therefore go to D." Each of these statements can be made equally well of a problem in geometry, a genealogy, an underground network, a fight between husband and wife, or the varnish painted on a canoe. Each can be said of every durable form (1.1.6). This is why "logic" is a branch of public works (1.4.4). We can no more drive a car on the subway than we can doubt the laws of Newton. *The reasons are the same in each case:* distant points have been linked by paths that were narrow at first and then were broadened and properly paved. By now nothing short of revolution or natural cataclysm would lead those who use these paths to suggest another route to the traveler. One logic is destroyed by another, in the way a bulldozer demolishes a shack. There is nothing miraculous about this displacement, though it can be dangerous if the expropriated avenge themselves.

2.4.7 The heterogeneous alliances that make certain strings of words coherent (2.1.8.0) form networks which may be very long and incommensurable—unless they choose to take each other's measure. "Can you doubt the link that joins B to C?" "No, I can't, unless I am ready to lose my health, my credit, or my wallet." "Can you loosen the bonds that tie D to E?" "Yes, but only with the power of gold, patience, and anger." The necessary and the contingent (1.1.5), the possible and the impossible, the hard and the soft (1.1.6), the real and the unreal (1.15.2)—they all grow in this way. For an entelechy there are only *stronger* and *weaker* interactions with which to make a world.

2.4.8 A sentence does not hold together because it is true, but *because it holds together* we say that it is "true." What does it hold on to? Many things. Why? Because it has tied its fate to anything at hand that is more solid than itself. As a result, no one can shake it loose without shaking everything else.

- Nothing more, you the religious; nothing less, you the relativists.

2.5.1 It is not good enough to be strongest; they also want to be best. It is never enough to have won; they also want to be right.

- "The strongest reason always yields to reasons of the strongest." This is the supplement of goodness that I would like to take away. The reasoning of the strongest is simply the strongest. "This world here below" would be very different if we were to take away this supplement, which does not exist, if we were to rob the victors of this little addition. For a start, it would no longer be a base world.

2.5.2 Power is the flame that leads us to confuse a force with those allies which render it strong (1.5.1). If we were to wear a welding mask, we could stare at the point of fusion without being blinded.

- I no longer wish to mistake the flash of a shield for the face of gray-eyed Athena, unless I wish to do so.

2.5.3 We can avoid being intimidated by those who appropriate words and claim to be "in power."

- On the night of the Sabbath the witches flew in potentia while their bodies slept on straw. No one believes this now, but the magic continues, the magic of those who believe they can travel *further* than their bodies and *beyond* the limits of their strength. The black Sabbath of the magicians of reason takes place every day of the week, and this magic has not yet encountered its skeptics (4.0.0).

2.5.4 We neither think nor reason. Rather, we *work* on fragile materials—texts, inscriptions, traces, or paints—with other people. These materials are associated or dissociated by courage and effort; they have no meaning, value, or coherence outside the narrow network that holds them together for a time. Certainly we can *extend* this network by recruiting other actors, and we can also *strengthen* it by enrolling more durable materials. However, we cannot abandon it even in our sleep.

- The butcher's trade extends as far as the practice of butchers, their stalls, their cold storage, their pastures, and their slaughterhouses. Next door to the butcher—at the grocer's, for example—there is no butchery. It is the same with psychoanalysis, theoretical physics, philosophy, accountancy, social security, in short all trades. However, *certain trades* claim that they are able to extend themselves potentially or "in theory" beyond the networks within which they practice. The butcher would never entertain the idea of reducing theoretical physics to the art of butchery, but the psychoanalyst claims to be able to reduce butchery to the murder of the father, and epistemologists happily talk of the "foundations of physics." Though all networks are the same size, arrogance is not equally distributed.

2.5.5 We cannot liberate ourselves from the powerful by means of "thought," but we will liberate ourselves from power when we have turned "thought" into work.

- The colloquial expressions we use for the work of thought (racking our brains, bending our minds, chewing over ideas) are not metaphors but point to the work of hands and bodies common to all trades. Why, then, is this trade of thought, unlike all others, held to be nonmanual? Because otherwise it would have to give up the privilege of going outside its networks. It would no longer be able to extend itself above the simple practice of tradesmen (2.1.7.2). Everyone prefers to set intellectuals apart (even if only to ridicule them) rather than to recognize that they work. Even if the believers do not benefit themselves from these free trips, they do not wish others to be deprived of the privilege of hovering outside time and space.

2.5.6 There is no difference between those who reduce, on the one hand, and those who want a supplement of soul, on the other. The two groups are the same. When they reduce everything to nothing, they feel that all the rest escapes them. They therefore seek to hold on to it with "symbols."

- The symbolic is the magic of those who have lost the world. It is the only way they have found to maintain "in addition" to "objective things" the "spiritual atmosphere" without which things would "only" be "natural."

2.5.6.1 We can be sure that whenever they talk of symbols, they are trying to travel without paying. They are hoping to move without leaving home, to link two actants with no trucks, no gas, and no freeway.

- Those who speak of "symbolic" behavior should be studied as magicians. They say that magic grasps through words what cannot be achieved by "efficacious practice." But this definition should be applied to *themselves*. Incapable of grasping forces through their trials, they invent "symbols" which cost and consume nothing "in addition to reality."

2.5.6.2 Since whatever resists is real, there can be no "symbolic" to add to "the real." Before having symbols "added " to them, actants lacked nothing. Thus, if we stop reducing them, this superfluous addition, in turn, becomes nothing.

- If only we were freed from the symbolic, the "real" would be returned to us. I am prepared to accept that fish may be gods, stars, or food, that fish may make me ill and play different roles in origin myths. They lead their lives, and we lead ours. Indeed, our lives have overlapped and made use of one another for so long that there are Jonahs in every whale, and whales in each of Melville's folios. Who will stop the translations of fishing, oceanography, diving— of everything that we and the fish use to take the measure of each other? That person is not yet born. (Interlude IV). Those who wish to *separate* the "symbolic" fish from its "real" counterpart should themselves be separated and confined (3.0.0).

2.5.6.3 We do not suffer from the lack of a soul. We suffer, on the contrary, from *too many* troubled souls that have never been offered a decent burial. They wander around in broad daylight like miserable ghosts. I want to exorcise these souls and persuade them to leave us alone with the living.

2.6.1 All research on foundations and origins is superficial, since it hopes to identify some entelechies which potentially contain the others. This is impossible. If we wish to be profound, we have to *follow* forces in their conspiracies and translations. We have to follow them, wherever they may go, and list their allies, however numerous and vulgar these may be.

- Those who look for foundations are reductionists by definition and proud of it. They are always trying to reduce the number of forces to one force from which the others can be derived. The greater their success, the more insignificant the chosen one becomes. The most profound is *also* the most superficial. We might just as well treat Queen Elizabeth as the United Kingdom, or the opening sentence (1.1.1) as the present text.

2.6.2 Those who try to possess what they do not have (1.5.1), to be where they are not, and to reduce what does not reduce are unfortunate, because they possess potency only potentially and have theory only in theory.

- We are now able to arrive at a moral of a less provisional kind (1.2.13). We will not try to pursue origins, to reduce practices to theories, theories to languages, languages to metalanguages, and so on in the way described in Interlude I. We will work with no more privilege or responsibility than anyone else, within narrow networks that cannot be reduced to others. Like everyone else, we will look for allies and openings, and sometimes we will find them. "This is not a very far-reaching moral, is it?" Quite so: *it does not get us very far*. It refuses to go in spirit to places where it is absent. When it moves, it pays its dues. We will no longer try to imitate Titan and carry the world on our shoulders, crushed by the infinite task of understanding, establishing, justifying, and explaining everything.

2.6.3 Because there is no literal or figurative meaning (2.2.2), no single use of a metaphor can dominate the other uses. Without propriety there is no impropriety. Each word is accurate and designates exactly the networks that it traces, digs, and travels over. Since no word reigns over the others, we are free to use all metaphors. We do not have to fear that one meaning is "true" and another "metaphorical." There is democracy, too, among words. We need this freedom to defeat potency.

2.6.4 How will we define this freedom to go from one domain to another, this scaling up of the networks, this surveying? Philosophy is the name of this trade, and the oldest traditions define philosophers as those who have no specific field, territory, or domain. Of course, we can do without either philosophy or philosophers, but then there might be no way to go from one province to the next, from one network to another.

2.6.5 There are only two ways of revealing forces. First, we can say that there are forces, on the one hand, and *other things*, on the other.

This amounts to denying the first principle (1.1.1). In this way "real" equivalences, "real" exchanges, and "real" essences are obtained, and the world is ordered by starting from masters (princes, principles, representatives, origins, foundations, causes, capital) and descending toward those who are dominated (inferred, explained, deduced, bought, produced, justified, caused). Second, we can uphold the first principle right to the end. If we do so, there are no longer any equivalences, reductions, or authorities unless the proper price is paid, and the work of domination is made public.

- The first way of working is religious in essence, monotheist by necessity, and Hegelian by method. It reduces the local to the universal and establishes potency. It abhors magic but nonetheless emulates its methods. The second way of working renders local what is local and deconstructs *potency*. It leads to skepticism about all magics, our own included.

Interlude III: Escaping from a Contradiction That, in the Author's Opinion, Might Have Perplexed the Reader

How can we say that nothing is by itself either reducible or irreducible (1.1.1) and then claim that there are nothing but trials of strength (1.1.2)? It is important to understand this paradox. If one thing can contain another—potentially, ideally, implicitly—there is truly something *more* than trials of strength: a supplement of soul, a living god, crowned princes or theories in charge of the world. *Certain* places become so much bigger than others that they include all the others "implicitly." They become impressive, majestic, sacred, intoxicating, dazzling, and thus bring with them all the impediments of terror. Those who believe it is possible to reduce one actor to another suddenly find themselves enriched by something that comes from beyond: beyond the facts, the law; beyond the world, the other world; beyond practice, theory; beyond the real, the possible, the objective, the symbolic. This is why reductionism and religion always go hand in hand: religious religion, political religion, scientific religion.

Of course, it is exciting to believe that one actor may contain the others because we start to believe that we "know" something, that there are equivalences, that there are deductions, that there is a master, that there is law and order. We have *two irons in the fire*, the real and the possible. In this way we become invincible, since we are able to make an attack "en double," like the witches of the Ivory Coast. A "trial of strength" can never be unfavorable to us, since even when we lose, we may still be right.

If we adopt the opposite principle and try to see how far we can get by denying the distinction, then we have to claim, by contrast, that nothing reduces to anything else. Yet, it will be said, things are linked together; they form lumps, bodies, machines, and groups. Of course this cannot be denied. But what kind of ties hold them together? Since there are no "natural" equivalences, these can be of only one kind: groping, testing, translating. As soon as the principle of irreducibility is accepted, it becomes necessary to admit this first reduction: that there is nothing *more than* trials of weakness. The distance between actors is never removed; neither is the distance between words. And if there are equivalences, then they have to be seen as problems, miracles, tasks, and costly results.

Thus there is no paradox. There are two consistent ways of talking. One permits reduction and builds the world by starting from potency. The other does not allow this initial reduction and thus manifests the work that is needed to dominate. The first approach is reductionist *and* religious; the second is irreductionist *and* irreligious.

Why should the second be preferred to the first? I still do not know, but I do not like power that burns far beyond the networks from which it comes. I do not like the verbiage, the exaggeration, and the saturation that leads to shortage of time and lack of breathing space. I would prefer to see the thin incandescent filament in all these flames, as if through the welder's mask. I want to reduce the reductionists, escort the powers back to the galleries and networks from where they came. I want to locate them in the gestures and the works that they use to extend themselves. I want to avoid granting them the potency that lets them dominate even in places they have never been.

If we choose the principle of reduction, it gives us plain, clean surfaces. But since there are many surfaces, they have to be ordered, and since they each occupy the whole of space, then they fight one another. It is necessary to survey their boundaries. Always summing up, reducing, limiting, appropriating, putting in hierarchies, repressing—what kind of life is that? It is suffocating. To escape, we have to eliminate almost everything, and whatever is left grows each day, like the barbarian hordes besieging Rome.

If we choose the principle of irreduction, we discover intertwined networks which sometimes join together but may interweave with each other without touching for centuries. There is enough room. There is empty space. Lots of empty space. There is no longer an above and a below. Nothing can be placed in a hierarchy. The activity of those who rank is made transparent and occupies little space. There is no more filling in between networks, and the work of those who do this padding takes up little room. There is no more totality, so nothing is left over. It seems to me that life is better this way.

Chapter 3
Anthropologics

3.1.1 How do things stand? What are the actants of which we speak? These entelechies, what do they want? They struggle to answer these questions themselves. To choose an answer is to strengthen one and weaken another.

- Every actant makes a whole world for itself (1.2.8). Who are we? What can we know? What can we hope for? The answers to these pompous questions define and modify their shapes and boundaries (1.1.6).

3.1.2 I don't know how things stand. I know neither who I am nor what I want, but *others* say they know on my behalf, others who define me, link me up, make me speak, interpret what I say, and enroll me. Whether I am a storm, a rat, a rock, a lake, a lion, a child, a worker, a gene, a slave, the unconscious, or a virus, they whisper to me, they suggest, they impose an interpretation of what I am and what I could be.

Interlude IV: Explaining Why Things-in-Themselves Get by Very Well without Any Help from Us

Things-in-themselves? But they're fine, thank you very much. And how are you? You complain about things that have not been honored by your vision? You feel that these things are lacking the illumination of your consciousness? But if you missed the galloping freedom of the zebras in the savannah this morning, then so much the worse for you; the zebras will not be sorry that you were not there, and in any case you would have tamed, killed, photographed, or studied them. Things in themselves lack nothing, just as Africa did not lack whites before their arrival. However, it is possible to force those who did perfectly well without you to come to regret that you are not there. Once things are reduced to nothing, they beg you to be conscious of them and ask you to colonize them. Their life hangs by no more than a thread, the thread of your attention. The spectacle of the world begins to turn around your consciousness. But who creates this spectacle? Crusoe on his island, Adam in his garden. How fortunate it is that you are there as saviors and name givers. Without you "the world," as you put it, would be reduced to nothing. You are the Zorros, the Tarzans, the Kants, the guardians of the widowed, and the protectors of orphaned things.

It is certainly hard work to have to extract the world from nothing every morning, aided only by the biceps and the transcendental ego. Crusoe gets bored and lonely on his island because of this drudgery. And at night, when you sleep, what becomes of the things that you have abandoned? You soon lose yourself in the jungle of the unconscious. Thus are your heroes doubly unhappy. Things-in-themselves muted and empty, expect from them their daily bread, while at night your heroes are powerless supermen who devour their own liver and leave their tasks undone.

What would happen if we were to assume instead that things left to themselves are lacking nothing? For instance, what about this tree, that others call *Wellingtonia*? Its strength and its opinions extend only as far as it does itself. It fills its world with gods of bark and demons of sap. If it is lacking anything, then it is most unlikely to be you. You who cut down woods are not the god of trees. The tree shows what it can do, and as it does so, it discovers what all the forces it welcomed can do. You laugh because I attribute too much cunning to it? Because you can fell it in five minutes with a chain saw? But don't laugh too soon. It is older than you. Your fathers made it speak long before you silenced it. Soon you may have no more fuel for your saw. Then the tree with its carboniferous allies may be able to sap *your* strength. So far it has neither lost nor won, for each defines the game and time span in which its gain or loss is to be measured.

We cannot deny that it is a force because we are mixed up with trees however far back we look. We have allied ourselves with them in endless ways. We cannot disentangle our bodies, our houses, our memories, our tools, and our myths from their knots, their bark, and their growth rings. You hesitate because I allow this tree to speak? But our language is leafy and we all move from the opera to the grave on planks and in boxes. If you don't want to take account of this, you should not have gotten involved with trees in the first place. You claim that you define the alliance? But this illusion is common to all those who dominate and who colonize. It is shared by idealists of every color and shape. You wave your contract about you and claim that the tree is joined to you in a "pure relationship of exploitation," that it is "mere stock." Pure object, pure slave, pure creature, the tree, you say, did not enter into a contract. But if you are mixed up with trees, how do you know they are not using you to achieve their dark designs?

Who told you that man was the shepherd of being? Many forces would like to be shepherd and to guide the others as they flock to their folds to be sheared and dipped. In any case there is no shepherd. There are too many of us, and we are too indecisive to join together into a single consciousness strong enough to silence all the other actors. Since you silence the things that you speak of, why don't you let them talk by themselves about whatever is on their minds, like grown-ups? Why are you so frightened? What are you hoping to save? Do you enjoy the double misery of Prometheus so much?

3.1.3 Those who speak always speak of others that do not speak themselves. They speak of him, of that, of us, of you . . . of who this is, what that wants, when the other happened. Those who speak relate to those of whom they speak in many ways. They act as spokesmen, translators, analysts, interpreters, haruspices, observers, journalists, soothsayers, sociologists, poets, representatives, parents, guardians, shepherds, lovers.

- Hobbes speaks of the "persona," the "mask," or the "actor" when he talks of those who speak on behalf of the silent. There are many masks, and they are not all known to the curators of Anthropology Museums.

3.1.4 Every actant decides who will speak and when. There are those it lets speak, those on behalf of whom it speaks, those it addresses. Finally, there are those who are made silent or who are allowed to communicate by gesture or symptom alone.

- Entelechies cannot be partitioned into "animate" and "inanimate," "human" and "nonhuman," "object" and "subject," for this division is one of the very ways in which one force may seduce others. We can make stone gods walk, deny the blacks a soul, speak in the name of whales, or make the Poles vote. Actors can always *be made* to *do* so, even though what they would do or say if they were left to their own devices is a mystery. (Probably they would not be "blacks," "whales," "Poles," or "gods" at all.)

3.1.5 A force is almost always surrounded by powers—by voices that speak on behalf of crowds that do not speak (1.5.0). These powers define, seduce, use, scheme, move, count, incorporate, and interrupt the force. Soon it is *no longer possible to distinguish* between (1.5.1) what the force says *itself,* what it *says* of itself, what the *powers* say it is, and what the *crowds* represented by these powers would have it say.

- By allying ourselves to words, to texts, to bronze, to steel, to places, or to emotions, we end up distinguishing shapes that can be classified, at least in peacetime. But these classifications never last for long before they are pillaged by other actors who lay things out quite differently.

3.1.6 Anything can be reduced to silence, and everything can be made to speak. Thus, any force may appeal to an inexhaustible *supply* of actors *who may be spoken for.*

- Ethnologists have shown us how ashes, curdled milk, smoke, ancestors, or wind may be made to talk. Their timidity has prevented them from seeing how others much closer to home make fossils, precipitates, blotting paper, genes, and tornadoes all talk. To be sure, psychoanalysts speak of the talkative "unconscious," but its repertoire is impoverished and it combines according to very few rules. In addition, psychoanalysts are prone to say that the subconscious has only "subjective" meaning. Yet all we need to do is read *The Times* to see how many more actors than the unconscious are made to speak in endless different ways: here legions of angels are mobilized to suppress vice; there thousands of pages of computer printout are generated to stop a nuclear plant; on the next page silent majorities are made to scream on behalf of the unborn child; a few pages earlier the dead were brought back to life to stop the desecration of a cemetary; on the back page whales had their spokesmen interrupt the deadly mission of a Japanese boat.

3.1.7 By definition *faithful* representatives cannot exist (2.2.1), since they say what their constituency has not said and speak in their place

(3.1.3). Every power can thus be *reduced to its simplest expression*. All that is needed is to have each of the actors in whose name the power speaks talk in turn. Then each actor will say what it wants itself, with neither censorship nor prompting. There is no quarter in the forces that reduce one another and call each other's bluff. "You speak in their name, but if I speak to them myself, what will they say to me?"

3.1.8 There is only one way in which an actor can prove its power. It has to make those in whose names it spoke *speak* and show that they all say the *same thing*. Once this is done, then the actor can say that it did not speak itself but faithfully "channeled" the views of others.

- A trade union organizes demonstrations by its members in the same way that Skinner's laboratory organizes demonstrations by rats. In each case the demonstrators and the rats have to be seen to be saying themselves the same as they have been made to say. And as for angels and devils, there are a thousand ways of finding signs of them—witnesses, stigmata, or prodigies—that will soften the hardened heart.

3.1.9 To make other forces speak, all we have to do is *lay them out* before whoever we are talking to. We have to make others believe that they are *deciphering* what the forces are saying rather than listening to what we are saying. Isn't this almost always possible?

- Elections, mass demonstrations, books, miracles, viscera laid open on the altar, viscera laid out on the operating table, figures, diagrams and plans, cries, monsters, exhibitions at the pillory—everything has been tried somewhere at one time or another in the attempt to offer proof.

3.1.10 Since a spokesman always says *something other* than do those it makes speak, and since it is always necessary to negotiate similarity and difference (1.2.1), there is *always room* for controversy about the fidelity of any interpretation. A force can always insinuate itself between the speaker and those that it makes speak. It can always make them say something else.

- The demonstrators did not say that they wanted the forty-hour week—they just attended in their thousands; the rats did not say that they had conditioned reflexes—they simply stiffened under electric shocks. Others can therefore intervene. The presence of the workers can be translated by saying that they were "paid by the union," and the stiffness of the rats can be interpreted as "an experimental artifact."

3.1.11 There is no *natural end* to such controversies. They may always be reopened (3.1.6). The only way to close them is to stop other actants from leading those that have been enrolled astray and turning them into traitors. In the end, interpretations are always stabilized by an array of *forces*.

3.1.12 A force becomes potent only if it *speaks for* others, if it can make those it silenced *speak* when called upon to demonstrate its strength, and if it can force those who challenged it to *confess* that indeed it was saying what its allies would have said.

- The trade union cannot stop its right-wing opponents from interpreting the demonstration differently. Skinner cannot prevent his "dear colleagues" from interpreting his experiment in other ways. If they could, they would certainly do so, but as it is, they can't. Others would ruin them if they tried.

3.2.1 What is the state of affairs? Where do things stand? What is the balance of forces? Using the multitudes which they make speak, some actants become powerful enough to define, briefly and locally, what it is all about. They divide actants, separate them into associations, designate entities, endow these entities with a will or a function, direct these wills or functions toward goals, decide how to determine that these goals have been achieved, and so on. Little by little they link everything together. Everything lends its strength to an entelechy that has no strength, and the whole is made "logical" and "consistent"—in other words *strong* (2.1.8).

- I am not trying to avoid giving an answer to the question, "What is the balance of forces?" Nevertheless, we must clear away the undergrowth so that *all* the answers will be able to display themselves.

3.2.2 None of the actants mobilized to secure an alliance stops acting on its own behalf (1.3.1, 1.3.4). They each carry on fomenting their own plots, forming their own groups, and serving other masters, wills, and functions.

- Forces are always rebellious (1.1.1); they lend themselves but do not give (1.5.1). This is true for the tree that springs up again, the locusts that devour the crops, the cancer that beats others at its own game, the mullahs who dissolve the Persian empire, the Zionists who loosen the hold of the mullahs, the concrete in the power station that cracks, the acryllic blues that consume other pigments, the lion that does not follow the predictions of the oracle—all of these have other goals and other destinies that cannot be *summed up*. The moment we turn our back, our closest friends enroll themselves under other banners.

3.2.3 How can those in whose name we speak be stopped from talking? How can those that have been recruited through good luck be cemented into a single block? How can the rebels and the dissidents be pacified? Is there a *single* entity anywhere that does not have to solve these problems? The answer is always the same, for there is only one source of strength: that which comes from joining together (1.3.2). But how can rebels be associated? *By finding more allies* which force the others to hold together, and so on, until a gradient of uncertain objects ends up making the first rank of the alliance resistant and thereby real (1.1.2).

- The notion of system is of no use to us, for a system is the end product of tinkering and not its point of departure (2.1.4). For a system to exist, entities must be clearly defined, whereas in practice this is never the case; functions must be clear, whereas most actors are uncertain whether they want to command or obey; the exchange of equivalents between entities or subsystems must be agreed, whereas everywhere there are disputes about the rate and direction of exchange. Systems do not exist, but systematizing is common enough; everywhere there are forces that oblige others to play the way they have always played (1.1.13).

3.2.4 As it associates elements together, every actor has a choice: to extend further, risking dissidence and dissociation, or to reinforce consistency and durability, but not go too far.

3.2.5 A well-defined state of affairs is the work of *many forces*. They agree about nothing and associate only via long networks in which they talk endlessly without being able to sum one another up. They intermingle, but they cannot reach outside themselves to take in what binds them, opposes them, and sums them up. However, despite everything, networks reinforce one another and resist destruction. Solid yet fragile, isolated yet interwoven, smooth yet twisted together, entelechies form strange fabrics. This is how we have imagined "traditional worlds," however far back we look.

- I do not talk of "culture," because the word has been reserved by Westerners to describe one of the detached entities used to constitute "man." Forces cannot be divided into the "human" and the "nonhuman." I do not talk of "society," because the associations that concern me are not limited to the few permitted by the "social." Again, I do not talk of "nature," because those who speak in the name of blood groups, chromosomes, water vapor, tectonic plates, or fish can only be temporarily and locally distinguished from those who speak in the name of blood, the dead, flood, hell, and fish. I would grant the term "unconscious" if we were sufficiently open-minded to designate things-in-themselves with it.

3.3.1 In order to spread far without losing coherence, an actant needs faithful allies who accept what they are told, identify themselves with its cause, carry out all the functions that are defined for them, and come to its aid without hesitation when they are summoned. The search for these ideal allies occupies the space and time of those who wish to be stronger than others. As soon as an actor has found a *somewhat more faithful* ally, it can force another ally to become *more faithful* in its turn. It creates a gradient that obliges the other allies to adopt a shape and retain it for the time being (1.1.12).

- We spend a lot of time looking for whatever happens to be harder in order to shape what is softer—a stone to serve as an anvil, a bioassay to measure the blood level of endorphine, a cow's tongue to let a virus penetrate the marrow, a law to curb the appetite of a lobby, a lobby to modify the law. The word "technology" is unsatisfactory because it has been limited for too long to the study of those lines of force that take the form of nuts and bolts.

3.3.2 If we want to stop forces from transforming themselves the moment we turn our back, we should avoid turning our back! Powers always dream of being everywhere, even when they are far away or long gone. How can they be present when other forces have pushed them to one side (1.2.5)? How can they extend themselves when everything localizes them? How can they be there and elsewhere, now and forever? Oh, the potency of the myth of potency! Anything that helps the present structure to last beyond the moment when force is withdrawn will do.

3.3.3   When a force has found allies that allow it to fix the ranks of other forces in a lasting manner, it can extend itself again. This is because the faithful are tied by such durable links that the force may withdraw without fear. Even when it is not there, everything will happen as if it were. In the end, there is simply a collection of forces which act for it but without it.

■ We sometimes call these machinations of forces "mechanisms." This term is poorly chosen—because it implies that all forces are mechanical, whereas most are not; because it emphasizes hardware at the expense of softer relations; and because it assumes that they are man-made and artificial, although their genealogy is precisely what is at stake.

3.3.3.1   To gain potency is always a matter of setting forces against one another. The power that results from the whole array is then attributed to the *last* force, trapped by all the others.

■ The reason I have talked of force from the outset should now be clear. It was not to *extend* technical metaphors to philosophy. On the contrary, the strength of machines or automatisms is achieved only rarely and locally. Only when we ignore all the other forces of which they are the *last in line* can we talk of "technology." The engine purring under the hood is only one of the possible forms taken by the conspiracy of forces. Diesel hoped to optimize the yield of social bodies as he had done for combustion engines. There was to be the same motor, the same research, the same optimization: compression, mixture, recovery, yield.

3.3.3.2   There is nothing special in these machinations apart from this Machiavellian injunction: collect the largest possible number of faithful allies that we can *inside,* and push those that we doubt to the *outside.* In this way we get a new division between the hard and the soft.

■ Those who are taken in by this division talk of "technology" and of "the social," without realizing that "the social" may be what is left over, like the shavings from the carpenter's plane. Every blueprint can be read as another *Prince:* tell me your tolerances, your benchmarks, your calibrations, the patents you have evaded and the equations you have chosen, and I will tell you who you are afraid of, who you hope will come to your support, who you decided to avoid or to ignore, and who you wish to dominate (Coutouzis: 1983).

3.3.4 Yet you cannot stop forces from playing against each other (3.2.2). There is no conspiracy, sorcery, logic, argument, or machine that can stop the mobilized actants from churning round and boiling as they search for other goals and alliances. The most impersonal machine is more crowded than a pond of fish.

- Contrary to Leibniz, in the movement of the watch there are also ponds full of fish and fish full of ponds. To be sure, it is always possible to find people who will say that machines are cold, impersonal, inhuman, or sterile. But look at the purest alloy: it is betrayed everywhere, too, like the rest of our alliances. Westerners always believe that motors are "pure" in the same way that arguments are "logical" and words are "literal." This is what the old captain said to Crusoe just before the shipwreck: "Beware of purity. It is the vitriol of the soul" (Interlude VI).

3.3.5 In order to extend itself, an actant must program other actants so that they are unable to betray it (3.3.3), despite the fact that they are bound to do so (3.3.4). There is only one way to resolve this quandary: since no individual link is solid, actants have to support one another; the moment numerous links are arrayed in tiers, they become reality.

- Since there is nothing but weakness, power is always an impression. However, this impression is all that is needed to change the shape of things by in*forming* or im*pressing* them. This is the mystery that has to be explained.

3.3.6 *We always misunderstand the strength of the strong.* Though people attribute it to the purity of an actant, it is invariably due to a tiered array of weaknesses.

Interlude V: Where We Learn with Great Delight That There Is No Such Thing as a Modern World

The whites were not right. They were not the strongest. When they landed on the island, their cannons only fired spasmodically and were no use at all in the face of poisoned arrows. Their engines were broken down more often than not and had to be repaired each day in a flood of grease and oaths. The Holy Book of their priests stayed as silent as the grave. The drugs of their doctors acted so erratically that it was scarcely possible to distinguish between their effects and those of medicinal herbs. Their books of law were beset with contradictions the moment they were applied to lineages or atolls. Each day the civil servants waited to be transferred or carried off by yellow fever. Their geographers were wrong about the names they gave to familiar places. Their ethnographers made fools of themselves with their blunders and their boorishness. Their merchants knew the worth of nothing and valued knickknacks, totems, wild pigs, and ground nuts equally. No, they were not the strongest, these uninitiated whites, racked by fever and smelling, acccording to the natives, of fish or rotten meat.

Yet they managed to make the island archaic, primitive, pagan, magic, precommercial, prelogical, pre anything we care to think of. And they, the whites, became in turn the "modern world."

This leads to the question that is asked on the shores of every ravaged country: how did such a rabble of weak, illogical, and vulgar nonbelievers manage to conquer the cohesive and well-policed multitudes? The answer to this question is simple. They were stronger than the strongest because they arrived *together*. No, better than that. They arrived *separately, each in his place and each with his purity*, like another plague on Egypt.

The priests spoke *only* of the Bible, and to this and this alone they attributed the success of their mission. The administrators, with their rules and regulations, attributed their success to their country's civilizing mission. The geographers spoke *only* of science and its advance. The merchants attributed all the virtues of their art to gold, to trade, and to the London Stock Exchange. The soldiers simply obeyed orders and interpreted everything they did in terms of the fatherland. The engineers attributed the efficacy of their machines to progress.

They believed in a separate order from which they drew their strengths. This is why they argued so much and distrusted one another. In their reports the administrators denounced the rapacity of the merchants. The learned found the proselytism of the priests scandalous, whereas the latter preached from the pulpits against the cruelty of the administrators and the atheism of the learned. The ethnologists despised everyone, while extracting their secrets and dragging their genealogies and myths from the natives one by one. They each believed themselves to be strong *because of their purity*—and indeed there were many worthy people who thought of nothing but the faith, the flag, philosophy, or finance.

Even so—and they knew this well—it was only because of the others that they were able to stay on the island at all. Since the priests were too weak to make God step out of the Bible, they needed soldiers and merchants to fill their churches. Since the merchants could not force the sale of totems with the strength of gold alone, they drafted priests and scientists to reduce their value. Since the scientists were too weak to dominate the island by science alone, they depended on police raids, forced labor, and the porters and interpreters lent to them by administrators.

Each group thus lent its strength to the others without admitting it, and therefore claimed to have retained its purity. Each went on attributing its strength to its domestic gods—gold, private convictions, justice, scientific rigor, rationality, machines, ledgers, or notebooks.

If they had come one at a time, they would have been overwhelmed by the island's inhabitants.

If they had come completely united, sharing the same beliefs and the same gods and mixing all the sources of potency like the conquerors of the past, they would have been *still more easily defeated*, since an injury to one would have been an injury to all.

But they came *together*, each one *separated* and *isolated* in his virtue, but *all supported* by the whole. With this infinitely fragile spider's web, they paralyzed all the other worlds, ensnared all the islands and singularities, and suffocated all the networks and fabrics.

Those who "invented the modern world" were not the strongest or the most correct, and neither are they today (Interlude VI).

3.4.1 How should we talk about all these things that hold together? Should we talk of economics, law, mechanisms, language games, society, nature, psychology, or a system that holds them all together?

- In James Bond films there is always a single black button that can undo the machinations of the evil genius, a button that the hero, disguised as a technician, manages to press at the end. Masked and disguised in white overalls, the philosopher reaches the point where extreme potency and extreme fragility coincide.

3.4.2 It is not a matter of *economics*. This makes use of equivalents, without knowing who makes equivalent, and of accountancy, without knowing who measures and counts. Economics always arrives *after* the instruments of measurement have been put in place—instruments that make it possible to measure values and enter into exchanges. Far from illuminating the trials of strength, economics disguises and represses them. At best it is a way of recording these trials once they have been stabilized.

- Once the instrument of measurement is established, we can do economics and calculate, economize, and save. In other words we can convince and enrich. But economists do not say how the instrument is established in the first place.

3.4.2.1   A general economy—a calculus of pleasure, genes, or profit—is not possible. It would need to reveal those who negotiate, those who have paid, those who have lost and won, how much the repayments are worth, and when the account should be closed.

3.4.3   It is not a matter of the *law*. This is a ratchet which, like any other (1.1.10), permits an actant to make the temporary occupation of a position irreversible. That which makes the law strong is not only texts but also the paralysis of those who dare not transgress what they believe to lie "potentially" in its scriptures; that is, the gap between law and force, or law and fact. If we wield this power, we can intimidate others and extend ourselves to new places no matter what the opposition. The strength of the law comes not from within it but from a poor despised rabble which gives it the force of fact: morals, words, truncheons, hopes, administrations, walls, telexes, files, finances, ulcers.

3.4.4   It is not a matter of *machines* or *mechanisms*. These have never existed without mechanics, inventors, financiers, and machinists. Machines are the concealed wishes of actants which have tamed forces so effectively that they no longer look like forces. The result is that the actants are obeyed, even when they are not there (3.3.3).

- Many people have dreamed of machines that can be extended to all relationships, but the dream is always haunted by a nightmare: a rebellion by sabotaging actants who lay traps for the smoothest-working machines. The strength of machines is drawn from other forces which come to be part of them—forces that others despise and repress; forces that are feebly associated, a vulgar rabble from the lower classes.

3.4.5   It is not a question of *language* or of language games (2.3.0, 2.4.3, 2.4.4). Words are not powerful but borrow their strength from compromises that are far removed from "belles lettres."

3.4.6   It is not a matter of *science*. If arguments were sovereign, they would have all the potency of a gouty monarch immured in a crumbling castle. If science grows, this is because it manages to convince dozens of actants of doubtful breeding to lend it their strength: rats, bacteria, industrialists, myths, gas, worms, special steels, passions, handbooks, workshops . . . a crowd of fools whose help is denied even while it is used.

- The superior school of facts is too often one of arrogance. Enlightenment leads to the crassest form of obscurantism.

3.4.7 It is not a matter of *society*. The meaning of the "social" continually shrinks—it has now been reduced to the level of "social" problems. It is what is left when everything else has been divided up among the powerful; whatever is neither economic, technical, legal, nor anything else is left to it. Do we really expect to bind everything together with this impoverished version of the social? Like a mayonnaise that does not take, it is bound to fail. The "social"—its actors, its groups, and its strategies—is too closely identified with human beings to pay heed to the feckless impurity and immorality of alliances.

- If *sociology* were (as its name suggests) the science of *associations* rather than the science of the social to which it was reduced in the nineteenth century, then perhaps we would be happy to call ourselves "sociologists."

3.4.8 It is not a matter of *intersubjective relationships*. Only in our day and age could we hope to find people so impoverished as to try to explain nuclear reactors, nation-states, or stock exchanges on the basis of "interactions." Psychology and its sister, psychoanalysis, think that they are rich in their infinite poverty. There is nothing to be said about this view except that it does not hold. The shrink who shrinks cannot expand to explain the rest (2.5.6.2).

- In the depths of the country there have always been retreats for people who want to make cathedrals out of matches or ball-point pens.

3.4.9 It is not a question of *nature* (3.2.5). Try to make sense of these series: sunspots, thalwegs, antibodies, carbon spectra; fish, trimmed hedges, desert scenery; "le petit pan de mur jaune," mountain landscapes in India ink, a forest of transepts; lions that the night turns into men, mother goddesses in ivory, totems of ebony.

See? We cannot reduce the number or heterogeneity of alliances in this way. *Natures* mingle with one another and with "us" so thoroughly that we cannot hope to separate them and discover clear, unique origins to their powers (Interlude IV).

3.4.10   It is not a question of *systems* (3.2.3). Since people know that the origin of power does not reside in the purity of forces, they locate it in a "system" of pure forces. This dream is always being reborn. Law is attached to economy, to biology, to language, to society, to cybernetics . . . Beautiful boxes are drawn, joined by nicely pointed arrows. Unfortunately for those who make systems, actors do not stand still for long enough to take a group photo; boxes overlap; arrows get twisted and torn; the law seeps into biology which diffuses into society. No, alliances are forged not *between* nice discrete parties but in a disorderly and promiscuous conflict that is horrible to those who worship purity.

3.5.1   We are always misunderstanding the efficacy of forces: we attribute things to them that they have only been lent (1.5.1). We hold them to be pure, though they would be completely impotent if this were the case. When we look at the way in which they work, we discover bits and pieces that can never be added up. Each network is sparse, empty, fragile, and heterogeneous. It becomes strong only if it spreads out and arrays weak allies.
- What can we compare with the weaknesses that make up a force? A macramé. Is there a knot that links men to men, neurons to neurons, or sheets of iron to sheets of iron? No. The rope of this Gordian knot has not yet been woven. But every day we see before our very eyes a macramé of strings of different colors, materials, origins, and lengths, from which we hang our most cherished goods.

3.5.2   Can we describe all networks in the same way? Yes, because there is no "modern world."
- For years ethnographers have said that it is impossible to study "primitive" or ancient peoples if we separate law, economy, religion, technology, and the rest. On the contrary, they have argued that these loosely linked mixtures may be understood only if we look very closely at places, families, circumstances, and networks. But when they talk of their own countries, they are committed to the separation of spheres and levels.

3.5.3 The "modern world" is the label on the button that unites extreme potency and extreme impotence (3.4.1). The heterogeneous and local application of weaknesses becomes a system of powers with prestigious names such as nature, economy, law, and technology.

- Like its zealots, those who abhor the modern world have invented more terms to describe it than the devout have found to celebrate the name of God. They say either "Vade retro, satanas" or "hear my prayer" to each of these invocations:

> the modern world
> secularization
> rationalization
> anonymity
> disenchantment
> mercantilism
> optimization
> dehumanization
> mechanization
> westernization
> capitalism
> industrialization
> postindustrialization
> technicalization
> intellectualization
> sterilization
> objectivization
> Americanization
> scientization
> consumer society
> one-dimensional society
> soulless society
> modern madness
> modern times
> progress

"Hear my prayer." "Vade retro, Santanas." Each of these words conceals the work done by forces and makes an anthropology of the here and now impossible. Yet it is really very simple: there is no modern world, or if there is one, it is simply a style, as when we say "modern style."

Interlude VI: In Which the Author, Losing His Temper, Claims That Reducers Are Traitors

I would like the following mystery to be explained. Why is it that since the Enlightenment we have delighted in talking of the "modern world"? Why is it that faith in the existence of this world links Althusser to Rockefeller, Zola to Burke, Sartre to Aron, and Lévi-Strauss to Hayek (1.4.6.2)? They say this "modern world" is *different* from all the others, absolutely and radically different. In the "modern world," but only there, the Being is not gathered by any being. This poor world is absolutely devoid of soul, and the tawdriest hand-carved clog has more being than a tin can. Why is it that we agree so easily with these premises *even before* we commit ourselves to "progress," "profit," or "revolution," or against "materialism," "rationalization," or "modernism"? Our most intelligent critics have done nothing for the last 150 years but complain of the damage caused by progress, the misdeeds of objectivity, the extension of market forces, the march of concrete in our towns, and dehumanization.

All right, let them complain, become indignant, criticize, and fight. This is necessary. But if they really want to win, why do they willingly hand over *the only thing* to the enemy that it needs to achieve domination?

The "enemies" on the fifty-eighth floor of the Chase Manhattan Bank, a quarter of a mile underground in the Red Army field marshall's bunker, three-quarters of a mile up in the spectrography room in the Mount Palomar Observatory, at four o'clock in the morning on the benches of the European Commission—"they" know very well that objectivization, rationalization, and optimization are pipe dreams that are about as accessible as the gates of Paradise. This is why every time they engineer a coup they are so surprised to discover that their enemies strike camp without engaging and leave the field of battle to them. In the face of the "modern world" everything flees. A captain of industry is not just a captain among many; he becomes a "capitalist." Such a radical discontinuity is created between him and his predecessors that those *who could have beaten him* run away. An engineer who, like the tinkerers, apprentices, and craftsmen of the past, brews up a slightly more favorable configuration of forces is converted into a Frankenstein by those who ought to be making an effort to prevent this configuration from turning into a monster.

For years we have *voluntarily* granted to the "modern world" a potency that it does not have. Perhaps once upon a time it bluffed and claimed superiority, but there was *no reason whatsoever* to concede this superiority (4.2.1). For too long the critics have withdrawn their troops from the Rhineland, intimidated by the rumbling of "rationalization" and "disenchantment." This massive strategic decision has left us *disarmed* in the face of the unmatched arrogance of captains of industry, technologists, and scientists. Munich was nothing compared with this unconditional surrender which grants the enemy everything that it would never have been able to win by itself. This pathetic melodrama by installments has been going on since the beginning of the nineteenth century. In the hope that this accusation will shame them, captains, engineers, and scholars are said to be rational and absolutely different. However, this simply crowns them with an accolade that they would never have won otherwise. Their opponents think themselves rich with what they have saved from the field of battle: the "spiritual," the "symbolic," the "warmth of interpersonal relationships," the "lived world," the "irrational," the "poetic," the "cultural," and the "past." We know the politics of the scorched earth, the politics of the worst case, but this strategy, which asks us to leave everything *untouched* and to flee, is new.

We witnessed these Munichs, though we could have fought and won. We saw this exodus in which the masses carried away their culture and poetry, though they lost everything in flight.

We must distrust those who believe in "true" market relationships, "true" equivalences, or "true" scientific deductions. No matter how polite, well-meaning, and cultivated they may be, they do not save the treasure that they claim to guard. In fact, they *disarm* those who might have the courage to approach the relations of force that create equivalences, machines, or knowledge. They weaken those who might, perhaps, have had *the strength to modify* that knowledge or those machines.

3.5.4 Fortunately, the world is no more disenchanted than it used to be, machines are no more polished, reasoning is no tighter, and exchanges are no better organized. How can we speak of a "modern world" when its efficacy depends upon idols: money, law, reason, nature, machines, organization, or linguistic structures? We have already used the word "magic" (2.1.11). Since the origins of the power of the "modern world" are misunderstood and efficacy is attributed to things that neither move nor speak, we may speak of magic once again (4.1.0).

3.5.5 What we are pleased to call "other cultures" have a number of secrets; ours may have only one. This is why "other cultures" seem mysterious to us and worth knowing, whereas our own seems both unknowable and stripped of mystery. This secret is the *only* thing that distinguishes our culture from the others: that it and it alone is not one culture among many. Our belief in the modern world arises from this denial. To avoid it, all we have to do is join together what we normally separate when talking of ourselves. We have to be the anthropologists of our own world.

3.6.1 What is it all about? What is the state of affairs? Someone speaks in the name of others who say nothing, and replies to my questions by putting me among the dumb. If the reply convinces me, I am no longer able to disentangle why, for it brings too many acolytes to support it.

3.6.2 Everything happens as if there were no trials of strength but rather a strange fantasy: "men" "discovering" "nature"!

3.6.3 Only in politics are people willing to talk of "trials of strength." Politicians are the scapegoats, the sacrificial lambs. We deride, despise, and hate them. We compete to denounce their venality and incompetence, their blinkered vision, their schemes and compromises, their failures, their pragmatism or lack of realism, their demagogy. Only in politics are trials of strength thought to define the shape of things (1.1.4). It is only politicians who are thought to be dishonest, who are held to grope in the dark.
- It takes something like courage to admit that we will *never do better* than a politician (1.2.1). We contrast his incompetence with the expertise of the well informed, the rigor of the scholar, the clairvoyance of the seer, the insight of the genius, the disinterestedness of the professional, the skill of the craftsman, the taste of the artist, the sound common sense of the ordinary man in the street, the flair of the Indian, the deftness of the cowboy who fires more quickly than his shadow, the perspective and balance of the superior intellectual. Yet no one does any better than the politician. Those others simply have somewhere to hide when they make their mistakes. They can go back and try again. Only the politician is limited to a single shot and has to shoot in public. I challenge anyone to do any better than this, to think any more accurately, or to see any further than the most myopic congressman (2.1.0, 4.2.0).

3.6.3.1 What we despise as political "mediocrity" is simply the collection of compromises that we force politicians to make on our behalf.
- If we despise politics we should despise ourselves. Péguy was wrong.

He should have said, "Everything starts with politics and, alas, degenerates into mysticism."

3.6.4 Someone speaks breathlessly to others who understand only what they want to hear. The story is about those who reveal themselves through enigmas and symptoms. From time to time those who are being talked about interrupt, furious that they have been betrayed. Sometimes those who were doing the talking stop, angry that they do not understand or have not been understood. Wavering, speakers grope from half-measure to compromise. They pick up forces which they test by trial and error and bind together into provisional alliances. When they like the result, they tie their fate to that of more durable materials. Little by little the forces grow, from combinations to arrangements, from one misunderstanding to the next, until the moment when others more numerous or skillful overwhelm them.

- Machiavelli and Spinoza, who are accused of political "cynicism," were the most generous of men. Those who believe that they can do better than a badly translated compromise between poorly connected forces always do *worse*.

3.6.5 Though it may sound strange, we are probably no more closely tied to most of the forces we speak for than a trade unionist is to the workers he represents, or a managing director is to his shareholders. I speak here of our dreams just as much as of our rats, our stomachs, or our machines.

- In the end politics is an acceptable model, so long as it is extended to the politics of things-in-themselves (4.5.0).

3.6.6 Worlds probably look more like a Rome than a computer. Or rather, the best-conceived computer should be thought of as a collage of displaced, reused ruins, a splendid Roman confusion (Kidder, 1981). Each entelechy looks like the court of Parma.

- Balzac said of Stendhal's *Charterhouse of Parma* that it was *The Prince* of the nineteenth century. Neither the secrets of the heart nor those of the court are grandiose—neither grandiose nor shabby, but irreducible, displaced, and betrayed.

Chapter 4

Irreduction of "the Sciences"

4.1.1 You can become strong only by association. But since this is always achieved through translation (1.3.2), the strength (1.5.1, 2.5.2) is attributed to potency, not to the allies responsible for holding things together (3.3.6). "Magic" is the offering of potency to the powerless. "They have eyes and see not, ears and hear not . . ."

- I have already talked of "magic." I used it first to deflate those who believe that they think (2.5.3) and then to treat all logics in the same way (2.1.11). I used it again in order to create an effect of symmetry between "primitive cultures," and "the modern world"(3.5.4). Now I want to use it to describe *all* errors about the origins of strength, *all* potency.

4.1.2 Do not trust those who analyze magic. They are usually magicians in search of revenge.

- My homage goes to Marc Augé who took the attack "en double" of the sorcerers of the Ivory Coast seriously (Augé: 1975). This greatly helped me not to take the attack "en double" by scientists seriously. When all magics are put on the same footing, we will have a new form of skepticism (Bloor:1976).

4.1.3 Conversely, once force is seen to lie in the alliance of weaknesses, potency vanishes. Of course, the forces are still there, but the illusion of potency is annihilated. Whatever displaces the magical impression of potency and escorts it firmly back to the network where it took form I call an "irreduction."

Be strong maybe, but potent never. Kill me, but do not expect me to wish for death and kneel before power. To force I will add *nothing*.

■ In Interlude III I said we should "reduce the reducers." In the old days the struggle against magic was called the "Enlightenment," but this image has backfired. The Enlightenment has since become the age of (ir)radiation. The head of the courageous researcher who tried to illuminate the shadows of obscurantism has since become the warhead of the missile that will blind us with light (Perhaps it is too late. Perhaps the missiles have already been launched. In this case, let us prepare for after the next war.)

4.1.4 When a network conceals its principle of association, I say that it displays "potency." When the array of weaknesses that makes it up is visible, I say that it displays "force."

4.1.5 We are suffering not from too little but from too much spirit. The *spirit*, alas, never lives up to the *letter*. Spirit is only a few words, among many to which the meaning of all the other words is unfairly attributed. Spirit thus becomes a potent illusion. Verily, I say unto you, the spirit is weak but the letter is willing.

■ When they speak, those who are religious put the cart before the horse. However, in practice they act quite differently. They claim that frescoes, stained glass windows, prayers, and genuflection are simply ways of approaching God, his distant reflection. Yet they have never stopped building churches and arranging bodies in order to create a focal point for the potency of the divine. The mystics know well that if all the elements that are said to be pointers are abandoned, then all that is left is the horrible night of Nada (1.4.6.1). A purely spiritual religion would rid us of the religious. To kill the letter is to kill the goose that lays the golden eggs.

4.1.6 What we call "science" is made up of a large array of elements whose power we prefer to attribute to a few.

- "Science" exists no more than "language" (2.4.3) or "the modern world" (3.5.2).

4.1.7 What we call "science" is chosen in a rather random manner from a motley crowd of actants. Though it represents the others, it denies this fact (3.4.6).
- Those who call themselves "scientists" always put the cart before the horse when they talk, though in practice they get things the right way round. They claim that laboratories, libraries, meetings, field notes, instruments, and texts are only *ways and means* of bringing the truth to light. But they never stop building laboratories, libraries, and instruments in order to create a focal point for the potency of truth. Rationalists know very well that if this subordinate material life were suppressed, they would be forced into silence. A purely scientific science would rid us of scientists. For this reason they are careful not to kill the goose that lays the golden eggs.

4.1.8 They are skeptical and unbelieving about witches and priests, but when it comes to science, they are credulous. They say without the slightest hesitation that its efficacy derives from its "method," "logic," "rigor," or "objectivity"(2.1.0). However, they make the same mistake about "science" as the shaman does when he attributes potency to his incantations. Belief in the existence of "science" has its reformers, but it does not have its skeptics, even less its agnostics.

4.1.9 Since nothing is by itself either reducible or irreducible to anything else (1.1.1), there cannot be tests and weaknesses on the one hand and *something else on the other* (1.1.2, 1.1.5.2, 2.3.4, 2.4.3, 2.5.1). However, the cunning of "science" (4.1.7) divides forces, making some seem strong while others look "true" or "reasonable."

4.1.10 If people did not believe in "science," there would be nothing but trials of strength. But even "in science" there are only trials of strength. This means that the irreduction of "science" is both necessary and difficult—necessary because it has become the *only obstacle* which stands in the way of our escaping from magic; difficult because it is our last illusion, and when we defend it, we believe that we are defending our most sacred inheritance.
- If this were not the case, I would not have to devote a whole chapter

to a critique of "science," for there is nothing very special about it.

Interlude VII: In Which We Learn Why This Précis Says Nothing Favorable about Epistemology

We would like to be able to escape from politics (3.6.3). We would like there to be, somewhere, a way of knowing and convincing which differs from compromise and tinkering: a way of knowing that does not depend upon a gathering of chance, impulse, and habit. We would like to be able to get away from the trials of strength and the chains of weakness. We would like to be able to read the original texts rather than translations, to see more clearly, and to listen to words less ambiguous than those of the Sibyl.

In the old days we imagined a world of gods where the harsh rules of compromise were not obeyed. But now this very world is seen as obscurantist and confused, contrasted with the exact and efficient world of the experts. "We are," we say, "immersed in the habits of the past by our parents, our priests, and our politicians. Yet there is a way of knowing and acting which escapes from this confusion, absolutely by its principles and progressively by its results: this is a method, a single method, that of 'science.' "

This is the way we have talked since Descartes, and there are few educated people on earth today who have not become Cartesian through having learned geometry, economics, accountancy, or thermodynamics. Everywhere we direct our best brains toward the extension of "science." It is with them that we lodge our greatest, indeed often our only, hopes. Nowhere more than in the evocation of this kingdom of knowledge do we create the impression that there is another transcendental world. It is only here that there is sanctuary. Politics has no rights here, and the laws that rule the other worlds are suspended. This extraterritorial status, available only to the "sciences," makes it possible for believers to dream, like the monks of Cluny, about reconquering the barbarians. "Why not rebuild this chaotic, badly organized world of compromise in accordance with the laws of our world?"

So what is this difference which, like Romulus and his plough, makes it possible to draw the *limes* that divide the scientific from other ways of knowing and convincing? A furrow, to be sure, an act of appropriation, an enclosure in the middle of nowhere, which follows up no "natural" frontier, an act of violence. Yes, it is another trial of strength which divides the forces putting might on one side and right on the other.

But surely this difference must represent something real since it is so radical, so total, and so absolute? Admittedly the credo of this religion is poor. All that it offers is *a tautology*. "To know" scientifically is to know "scientifically." Epistemology is nothing but the untiring affirmation of this tautology. Abandon everything; believe in nothing except this: there is a scientific way of knowing, *and other ways,* such as, the "natural," the "social," or the "magical." All the failings of epistemology—its scorn of history, its rejection of empirical analysis, its pharisaic fear of impurity—are its only qualities, the qualities that are sought for in a frontier guard. Yes, in epistemology belief is reduced to its simplest expression, but this very simplicity brings success because it can spread easily, aided by neither priest nor seminary.

Of course, I am exaggerating. The faith has some kind of content. Technically, it is the negation of the paragraph with which I started this précis (1.1.2). Since the gods were destroyed, this faith has become the main obstacle that stands in the way of understanding the principle of irreduction. Its only function is passionately to deny that there *are* only trials of strength. "Be instant in season, out of season," to say that "there is something in addition, there is also reason." This cry of the faithful conceals the violence that it perpetrates, the violence of *forcing* this division.

All of which is to say that this précis, which prepares the way for the analysis of science and technology, is not epistemology, not at all.

4.2.1 "Science"—in quotation marks—does not exist. It is the name that has been pasted onto certain sections of certain networks, associations that are so sparse and fragile that they would have escaped attention altogether if everything had not been attributed to them.

- Two to three percent of the GNP of a few industrial nations, two-thirds of which is spent on industry and for military purposes—that is not much. The tiny fraction that remains is valued by a few thousand people only, communicated to a few thousand more, and popularized for the benefit of a few million brave souls who hardly understand it at all. For billions of others all these networks are *invisible*.

4.2.2 "Science" has no standing of its own. It takes shape only by denying what carried it to power and by attributing its solidity not to what holds but to what is held together (2.4.7). With this denial "it" ignores even itself.

- If the mongrel tribes that do the dirty work were withheld from "physics," its elucubrations could not be distinguished from those of alchemists or psychoanalysts: was this possible in the past when there were not so many tribes?

4.2.3 "Science" is an artificial entity separated from heterogeneous networks by *unjust means*. There are two measures, one for the "scientists" and the other for the rest.

- If a capitalist sells an unprofitable factory, he is accused of rapacity.

But if an illustrious scientist renounces a discredited hypothesis, then on the contrary he is held to be showing disinterestedness. If an unfortunate witch attributes success in battle to a magic rite, she is mocked for her credulity. But if a celebrated researcher attributes the success of her laboratory to a revolutionary idea, no one laughs, even though everyone should. The thought of making a revolution with ideas! If consumers cut their steak into small pieces to make it easier to chew, no one comments. But if a famous philosopher in Amsterdam asserts that we must "divide up each of the difficulties into as many parts as possible," no greater admiration could be expressed for "a method of rightly conducting the reason and seeking for truth in the sciences." If the most obscure Popperian zealot talks of "falsification," people are ready to see a profound mystery. But if a window cleaner moves his head to see whether the smear he wants to clean is on the inside or outside, no one marvels. If a young couple move a piece of furniture in their living room and conclude, little by little, that it does not look right and that all the furniture will have to be moved for everything to fit again, who finds this worthy of note? But if "theories" rather than tables are moved, then people talk excitedly of a Kuhnian "paradigm shift." I am vulgar, but this is essential in a domain where injustice is so profound. They laugh at those who believe in levitation but claim, without being contradicted, that theories can raise the world.

4.2.4 "Science" only gives the impression of existing by turning its existence into a *permanent miracle*. Unable to admit its true allies, it is forced to explain one marvel with another, and that one with a third. It goes on until it looks just like a fairy tale.

■ Some say that it is a miracle that "mathematics is applicable to physical reality." Others say that "the most incomprehensible thing about the universe is that it's at all comprehensible." Still others express amazement that the laws of physics "are universally applicable," that Newton discovered them, and that Einstein revolutionized them. "Science" becomes truly a circus sideshow with geniuses, revolutions, and dei ex machina. But no one talks of the chamber of horrors down below. When we become agnostic, we have to admit that most places of scientific pilgrimage look much like Lourdes, but more gullible still, for they mock Lourdes!

4.2.5 "Science" is a sanctuary only so long as we treat the winners and the losers asymmetrically.

- Nobody can separate the "internal" history of science from the "external" history of its allies. The former does not count as history at all. At best it is court historiography, at worst the Legends of the Saints. The latter is not the history of "science," it is history.

4.2.6 Belief in the existence of "science" is the effect of exaggeration, injustice, asymmetry, ignorance, credulity, and denial. If "science" is distinct from the rest, then it is the end result of a long line of coups de force.

4.3.1 "Science" is much too ramshackle to talk about. We must speak instead of *the allies* which certain networks use to make themselves stronger than others (1.3.1, 2.4.1, 3.3.1). In this way we will see force instead of potency (4.1.5).

4.3.2 Knowledge does not exist—what would it be (1.4.3)? There is only know-how. In other words, there are crafts and trades. Despite all claims to the contrary, crafts hold the key to knowledge. They make it possible to return "science" to the networks from which it came (Introduction).

4.3.3 We do not think. We do not have ideas (2.5.4). Rather there is the action of *writing*, an action which involves working with *inscriptions* that have been extracted; an action that is practiced through *talking* to other people who likewise write, inscribe, talk, and live in similarly unusual places; an action that *convinces* or fails to convince with inscriptions which are made to speak, to write, and to be read (3.1.0, 3.1.9).

- When we talk of "thought," even the most skeptical lose their critical faculties. Like vulgar sorcerers, they let "thought" travel like magic at high speed over great distances. I do not know anyone who is not credulous when it comes to ideas. Yet "thought" is really quite simple, for when we write about other inscriptions, we actually cover great distances in a few centimeters. Maps, diagrams, columns, photographs, spectrographs—these are the materials that are forgotten, the materials that are used to make "thought" intangible.

4.3.4 Despite all impressions to the contrary, standing by what is written on a sheet of paper alone is a risky trade. However this trade is no more miraculous than that of the painter, the seaman, the tightrope walker, or the banker.

- It is interesting to see the Greek leaning over the blinding surface of the parchment and obsessively following the incisions of the stylus, even when these lead to sophisms. It is fascinating to see the Church Fathers spreading out the different versions of the same text and learning to ply the trade of exegesis, the mother of *all* scientific disciplines. It is stimulating to follow the Italian as he rewrites the book of nature in mathematical form in his *Dialogues* (Eisenstein: 1975). It is fascinating to study, as I did for two years, the needles that scratch the drums of physiographs; to see how traps are set to make the things that are talked about write (3.1.5) and speak directly to those whom one wishes to convince. These bizarre texts, which are not sacred writings but inscriptions produced by rat viscera or the open hearts of dogs, are strangely alluring. They are all very beautiful, I agree. They represent a lot of work and much dexterity, but they are not miraculous. There is nothing immaterial in the endless breaking of bindings, clicking of pens, clattering of daisy wheels, and scratching of styli. There is nothing immaterial about this obsession with writing, inscription, diagrams, and spectra.

4.3.5 Do they turn toward nature? What could this mean? Look at them! They lean over their writing and talk to one another *inside* their laboratories. Look at them! Their only principle of reality is one that they have determined themselves (1.2.7). Look at them! The "external" referents they created exist only *inside* their world (1.2.7.1).

4.4.1 Whatever is local always stays that way. No kind of work is *more* local than any other unless it has been conquered (1.2.4) and forced to yield a trace. Then it can be worked on *in its absence*.

- An African hunter who covers dozens of square miles and who has learned to recognize *hundreds of thousands* of signs and marks is called a "local." But a cartographer who has learned to recognize a *few hundred* signs and indices while leaning over a few square yards of maps and aerial photographs is said to be more universal than the hunter and to have a global vision. Which one would be more lost in the territory of the other? Unless we follow the long history that has turned the hunter into a slave and the mapmaker into a master, we can have no answer to this question. There is no pathway between the local and the global because there *is* no global. Instead we have geographers, planes, maps, and International Geodesic Years.

4.4.2 "General ideas" can be built, but to do so is no more and no less difficult than building a railroad network. We have to pay for a "general idea." We cannot move from one table to another via the concept of "table." To move, we need a network as expensive to maintain as a railroad system, with its shunters, its striking railroadmen, its accountants, and its signals.

- Scholars understand the principle of the "privatization of benefits, the nationalization of losses" very well. They lead us to believe that they think and that ideas are free, but then they ask us to pay for their laboratories, their lecture theaters, and their libraries (4.1.9).

4.4.3 When a series of locations has been mastered and joined together in a network, it is possible to move from one place to another without noticing the work that links them together. *One* location seems "potentially" to contain all the others. I am happy to call the jargon used to get by inside these networks "theory," as long as it is understood that this is like the signposts and labels that we use to find our *way back*.

- There is the jargon of Phoenician coastal traders, of longshoremen, of financiers, of white-coated people who count in light-years and weigh things by the picogram. How can they all understand one another? They do not have the same destinations. Nor do they move along the same lines of force or manipulate the same traces. What we call "theory" is no more and no less real than a subway map in the subway (2.1.7)

4.4.4 "Universality" is as local as the rest. Universality exists only "in potentia." In other words it does not exist unless we are prepared to pay the high price of building and maintaining costly and dangerous liaisons.

- If everything happens locally and only once (1.2.1) and if one place cannot be reduced to another, then how can one place contain another? Do not accuse me of nominalism. All the parts of an army *may* be linked to a headquarters. The officers of the Strategic Air Command *may* work on a map of the world that measures three meters by four. All the clocks in the world *may* be synchronized if a universal time is built. I simply want the cost of creating these universals and the narrow circuits along which they run to be added to the bill.

4.4.5 So you believe that the application of mathematics to the physical world is a miracle? If so, then I invite you to admire another miracle; I can travel around the world with my American Express card. You say of the second, "That's just a network. If you step out of it by so much as an inch, your card will be valueless." Quite so. This is what I am saying about mathematics and science, *nothing more and nothing less.*

- The second-degree equation has an area of diffusion that can be mapped like everything else. Its invention, translation, and incorporation into other practices may be followed in the same way that we document the spread of the harness, the stern-mounted rudder, the bow tie, the clock escapement, or intelligence tests. But we cannot resist separating trades into two heaps. Some are firmly embedded in their contexts, while others float like spirits out of context. I want to bury those spirits at the bottom of their networks to stop them from returning after dark to haunt us.

4.4.5.1 The "universal" can no more swallow the particular than historical paintings can replace still lifes. Theories cannot be abstract, or if they are, the name refers to a style, like abstract painting.

- When someone talks to me about a universal, I always ask what size it is, and who is projecting it onto what screen. I also ask how many people maintain it and how much it costs to pay them. I know that this is in bad taste, but the king is naked and seems to be clothed only because we believe in the universal.

4.4.6 How are "abstraction," "formalism," "exactness," and "purity" achieved? Like cheese, by filtering, seeding, molding, and aging. Or like petrol, by refining, cracking, and distilling. We need dairies and refineries. These are all expensive processes, impure crafts that smell.

4.4.6.1 The *work* of abstraction is no more abstract than the work of the engraver; the *trade* of the formalizer is no more formal than that of the butcher; the work of purification is no more pure than that of the sanitary inspector. To say that some procedures are pure, formal, or abstract is to confuse a verb with an adjective. We might as well say that tanning is tanned, filtering is filtered, or logic is logical.

4.4.7　It is no more in our power to be abstract than to talk properly (2.2.1).

4.4.8　Networks are tenuous, fragile, and sparse. We read and we write inside them. We are able to convince only by extending the network, in other words by reducing the scale of whatever is absorbed. The result is that a few people, sitting around a table in a single room, can survey everything. *What could be simpler?* There is nothing here to make a fuss about.

Interlude VIII: In Which a Little Bit of Everyday Sociology Shows What Measures Are

The butcher used his scale, and I paid 25 francs. I did not try to bargain with him because the prices were on little tags stuck in the meat. He had decided on the price per kilo after he returned from the wholesale market and read his trade newspaper. When I left the shop, I took the No. 80 bus. I knew it was the 80 because the number was clearly displayed on the front of the bus. When the driver heard the signal from the bus company headquarters, he set off. This signal was relayed from the speaking clock of the Observatory of Paris, which was linked in its turn to the network of atomic clocks that harmonize time. I was not afraid of the ticket inspector. I had my bus pass, so the inspector checked my photograph, said "thank you" politely, and moved on. When I arrived at the Institute, I put my magnetic card into the electronic timekeeping clock which keeps track of the number of hours that we put in and their spread. There had been an argument with the unions about this clock for four years. Finally an agreement was reached, thanks to collective bargaining with The National Association of the Workers of the Proof (NAWOP). I still have fifteen hours to do this week.

After hanging up my coat, I went straight to see how my cells were getting on. The colonies had become quite visible. I counted the spots that they had made on the gel and wrote the results down in two columns in my laboratory notebook—a fine book, leather bound, just like my father's account book. I discussed yesterday's results with Dietrich, but his peaks are much clearer. He claimed that his neurotransmitter was a hundred times more active than mine, but I told him that we could argue for hours about it because he had not got enough points to draw a curve. Dietrich is still young. He is always jumping to conclusions. We talked for several hours. Finally he accepted that I was not going to use his work in my article. I do not want to weaken it and have people jumping on it saying that the results do not stand up. I want the article to be beyond reproach and accepted immediately by the referees of *Endocrinology*. Dietrich took my refusal badly. He looked quite crestfallen. He is too easily discouraged for this line of work. Fortunately he then bucked up and decided to do another series of rats in order to strengthen his data. If these turn out to be solid then I will use them. They might even reinforce my point, in which case I will make him junior co-author. This would not weaken my position.

In the canteen we discussed the forthcoming elections. As long as there are only opinion polls, we can argue about the relative position of the Socialists and the Communists until the cows come home. These polls do not count. Like Dietrich's rats, their samples are too small. What is needed is a truly grand experiment in which all the votes are counted and everyone can see that everything is aboveboard. Only then will we know whether the Communists are two percent weaker then the Socialists.

Then Brunel came along, and we chatted. He is an economist, and we are always teasing him because he claims to be a scientist. He admitted that we cannot tell whether the money supply is increasing or decreasing. There have been three meetings in his department to decide whether or not to include the discount bills between banks, or something like that. It seems that taking out a single line of the calculation makes it possible to change the results completely and prove that we have beaten inflation. It is quite incredible, but as Brunel replied when we teased him about it, "You've got more rats around than I've got economies."

Even so, our rats are threatened, as we discovered after lunch when we met in the lab. We have to find half a million dollars to pay for all the rats that we need for our fifty articles a year. We will manage somehow.

In the evening I met Adele. She had made a mistake when she was taking her temperature and was worried. We discussed her temperature curve for half an hour, but in the end she accepted my word because I am always taking the temperatures of rats. I told her that it would be more sensible to take the pill, as I do. Then all you have to do is to read which day it is and take the right pill. You cannot make a mistake; it is as simple as a bus pass. Then we went home to meet our men. The moment Adele told them she was worried, they wanted to persuade her that really her subconscious was speaking. She does not know much about it, but they sounded as if they knew what they were talking about. Finally she gave up. As she put it, "You're not allowed to have private problems. You've got to lie on the psychiatrist's couch and discuss them." Henry told her that she was just the same with her cosmology. Yesterday she went on for hours trying to persuade us that the Big Bang was a load of nonsense. Henry had said that perhaps there were several cosmologies. She thought this was nonsense.

It was late, so we went home. We argued for a quarter of an hour with the driver, who wanted to add ten percent to the meter reading. I told him I would never do that with my rats, add ten percent for no reason. He replied that it was a unilateral decision taken by the taxi owner's association which was in dispute with the City Council. We end up paying either way. In the mail I found my pay check with a further one percent deducted with the agreement of the union to pay for an increased social security levy. I set my clock and checked the alarm three times so that I would not worry all night that I might wake up late.

4.5.1 In scientific trades, as in all others, we learn how to increase our force locally (Part One).

4.5.2 The supplement of force gained in the laboratory comes from the fact that lots of small objects are manipulated many times, that these microevents can be recorded, that they can be reread at will, and that the whole process can be written for people to read. Skill is needed and lots of money, but witchcraft is not involved.

- It does not matter whether they are nebulas, corals, lasers, microbes, Gross National Products, or I.Q. scores. It does not matter whether they are "infinitely large" or "infinitely small." They are only talked about *with confidence* when they are brought to a small space where they can be dominated by a few people and made to display signs—curves, figures, points, rays, or bands—which are so simple that agreement is possible. We can only stutter about the rest.

4.5.2.1 The rule is quite simple: if we want to increase our strength, use a thousand against one on topics that will pay a hundred to one.

- Imagine an anthrax bacillus which has lived for millions of years hidden in the crowd of its cousins. One day it finds itself alone with its children under the blinding light of a microscope that is dominated by Pasteur's enormous beard. It has nothing to live on but urine (Part One). This is a good example of a reversal in the balance of forces. Doesn't exactness always grow out of such reversals? It really requires the blindness of faith to ignore the trials of strength that take place in the torture chambers of science—bioassays, tensimeters, linear accelerators, presses, needles, styluses, vacuum pumps, calorimeters. To remain blind in the face of those trials is what "courageously resisting the *question*" really amounts to! Those who believe in "science" in spite of this are the real martyrs.

4.5.3 So they are more certain of themselves than others are? Of course they are! They have tried their arguments out dozens of times on small-scale models and made all possible mistakes. Obviously they are more certain than those who only have one go.

- The respected expert is indistinguishable from the politician who is scorned by everyone. The expert makes large numbers of secret small-scale mistakes and confidently emerges from hiding *at the end of the day*. The politician makes really grand mistakes and has to perform in front of everyone. Here the decisions are made— *before* the mistakes (3.6.3). All people are the same—equally honest, equally erratic. How could they possibly be otherwise?

4.5.4 The only way to be strong again is to reproduce relations of force that were once favorable. *There is no such thing as prediction.* Prediction is the repetition of something that has already taken place, scaled up or scaled down. Only magicians believe that they can foretell the future.

- If we find miraculous the fact that unvaccinated sheep die at Pouilly- le-Fort or that Voyager II passes through the rings of Saturn at the prescribed moment, then we should find Hamlet's death in the last act equally amazing. No prediction is more than stage management, learning how to repeat the dress rehearsal—though this does not prevent stage fright and suspense. As far as forecasts are concerned, Pasteur, Shakespeare, and NASA are indistinguishable. If they had to improvise or predict, they would jabber incoherently like the Pythia, just as we do when we leave the shelter of our trades. And Shakespeare would probably be less incoherent than any of the others. In the theater of proof, or in the theater, plain and simple, all directors are the same, equally erratic and equally honest. How could they be different?

4.5.5 The only way to know is through trials of strength. "Knowledge" is the state of this battlefront. It extends no further. How could it? (1.1.0).

- The scientists say that they reach conclusions in the laboratory, "everything else being equal," but then they forget, preferring to travel by magic to other places and legislating as if they were still home.

4.5.6   Nothing can be known outside the networks organized and manipulated by know-how (1.3.7), but those networks may be extended.

4.5.7   There is no such thing as "knowledge" (4.3.2), but it is possible to realize, that is, to make real, to understand.

- The mystery of *adequatio rei et intellectus* is simply the extension of the laboratory. If we do not believe in magic, this extension is visible, but if we convert an array of weaknesses into a miraculous power, this extension is concealed. "Science" has no outside (4.3.5), but only narrow galleries which allow laboratories to extend and insinuate themselves into places that may be far away.

4.5.7.1   Nothing escapes from a network, least of all know-how, but who doubts that a network which pays the price can extend itself? "Prove to me that this substance which works so well in Paris is equally good in the suburbs of Timbuktu."
"But what on earth for? There is a universal law."
"I don't want to have to *believe* in it. I want to *see* it."
"Just wait until I have built a laboratory, and I'll prove it to you . . ."
A few years and a few million dollars later in the brand-new laboratory I see the proof that I asked for with my own eyes. I step away, travel a few miles, and pose the question again:
"Prove to me that . . ."

- When people say that knowledge is "universally true," we must understand that it is like railroads, which are found everywhere in the world but only to a limited extent. To shift to claiming that locomotives can move beyond their narrow and expensive rails is another matter. Yet magicians try to dazzle us with "universal laws" which they claim to be valid even in the gaps between the networks!

4.5.7.2.   How can know-how be extended? Like radios that are made in Hong Kong, or multiplication tables! There must be buyers and sellers, teachers and commercial circuits, representatives and books that are held to be authoritative.

- We say that the laws of Newton may be found in Gabon and that this is quite remarkable since that is a long way from England. But I have seen Lepetit camemberts in the supermarkets of California. This is also quite remarkable, since Lisieux is a long way from Los Angeles. Either there are two miracles that have to be admired together in the same way, or there are none.

4.5.7.3 People usually talk of "scientific truth" in hushed tones. But there have ever been only three ways of celebrating it: consistency—"it is logical"; representation—"it fits"; efficacy—"it works." These three expressions simply serve to indicate the extent to which a network has expanded.

- In kitchen Latin we would say *adequatio laboratii et laboratorii; adequatio laboratorii et alius laboratis, adequatio laboratorii et vulgi pecoris.*

4.5.8 One form of know-how is no more "true" than another. It is neither more nor less true than a coffeepot, a tree, or a child's face. There they are, a momentarily stable line of forces (1.1.6). The word "true" is a supplement added to certain trials of strength to dazzle those who might still question them.

- Rationalists laugh at the ordeal which makes the victor in combat right. However, each day they crown the victors in scientific controversies by claiming that they have purer hearts and more rational minds! There are two measures, two standards (4.2.3).

4.5.9 We can say that whatever resists is real (1.1.5). The word "truth" adds only a little supplement to a trial of strength. It is not much, but it gives an impression of potency (2.5.2), which saves what might give way from being tested.

- Relativists and idealists have never been able to hold their postures for long (1.3.6), for the statements that come from laboratories stand up, resist, and are thus real (2.4.7). But they are right: this is no reason to believe in fairy tales.

4.5.10 If something resists, it creates the optical illusion among those who test it that there is an object that can be seen and described causing this resistance. But the object is an effect, not a cause. The illusion disappears when the battlefront moves and discreetly reappears as soon as the battlefront stabilizes again.

- "Real worlds out there" are the consequences of lines of stable force and not the cause of their stabilization.

4.5.11   We can perform, transform, deform, and thereby form and inform ourselves, but we cannot *describe* anything. In other words there is no representation, except in the theatrical or political senses of the term.

- The difficulty with the "sciences" perhaps arises from the fact that work with the hands brings inscriptions that are read by the eyes. Perhaps epistemology is a confusion of the senses. We follow the dazzled gaze but forget the hands that write, combine, and mount. But there is no "theory," no "contemplation," no "speculation," no "prevision," no "vision," and no "knowledge." Plato's sun neither burns nor turns in the sky. But inside the networks there are electrons, light bulbs, and projectors which consume electricity and are objects like anything else. Such lamps are not surrounded by a halo of mystery. They are plugged into their sockets by real hands.

4.6.1   Why should we be surprised that those who amass a surplus of force and add their own weight to a conflict where no one has the upper hand should win?

4.6.2   When we cannot win with our own forces alone, we talk of those whom we command as "power," and the balance as "knowledge." Our opponents may be able to resist the addition of "forces," but not the superiority of "knowledge" over "power."

- This is how the division between forces that we have met from the beginning may be explained (1.1.5.2, 4.1.9). The distinction refers not to anything obvious but to a strategem that multiplies additional forces tenfold by describing some of them as "science." "Science" is like the sword of Brennus thrown into the balance. Yes, *vae victis,* for they will be called "illogical," "bad," and "unreasonable." "From him that hath not shall be taken away even that which he hath."

4.6.2.1   If it were possible to explain "science" in terms of "politics," there would be no sciences, since they are developed precisely in order to find other allies, new resources, and fresh troops.

- This is why the sociology of science is so congenitally weak. Auguste Comte, the father of scientism and sociology, has invented a fancy system of double-entry bookkeeping. Science is not politics. It is politics by other means. But people object that "science does not reduce to power." Precisely. It does not reduce to power. It offers *other means*. But it will be objected again that "by their nature, these means cannot be foreseen." Precisely. If they were foreseeable, they would already have been used by an opposing power. What could be better than a fresh form of power that no one knows how to use? Call up the reserves! Homage to Shapin and Schaffer (1985).

4.6.3 Now that we are no longer fooled by these maneuvers, we see spokesmen (3.1.3), whoever they may be, speaking on behalf of other actors, whatever they may be. We see them throwing their ranks of allies, some reluctant, some bellicose, into battle one after the other.

- The first advances, followed by his microbes; the second, by his angry workers, a third, by his whales whose numbers and needs he knows and which he wants to save; the fourth, by his batallions; the fifth, by his Koran and petrodollars; the sixth, by the great interests that he represents; the seventh, by a bulldozer; yet another, by his sheep and his dog. They are all arrayed and ranked according to the numbers they have enrolled. They all establish what is real in the battlefront of their trials. If we try to divide this crowd into the human and the nonhuman or into the "political" and the "scientific," then we are making a mistake—we are, I insist, committing treason (4.7.0)

4.6.4 But what would the entelechies that have been enrolled in our conflicts say if they could speak for *themselves?* The "same" as they are made to say. When brilliant demonstrations force us to confess this every day, how can there be any doubt?

- Sometimes people talk of "nature" when referring to the crowd of slaves and subjugated actants that have been reduced to silence, or when speaking about the commands given by a class of researchers who, in turn, march to the tune of a handful of "great thinkers." But it is most unlikely that the forces are really like this. After all, only two to three percent of the GNP of a few countries circulates inside the sparse and fragile networks of "science." We might as well try to reduce all the journeys in the world to airline networks (2.1.8.1). An actor must have achieved hegemony to speak in the singular of "nature" or "the real world out there." Hegemony is the cause of "the world" in the singular, not its consequence.

4.6.5 But what would the innumerable actants enrolled in our conflicts and our brilliant demonstrations say if they were able to talk for themselves? We have no idea. Not because they are unknowable (1.2.12), nor because they are ineffable (2.2.3), but because no one has ever tried, or rather because those who have tried have returned weaker than when they left.

4.6.6 We still know very little "objectively." We only know anything because some forces grow at the expense of others. We do not have the slightest idea about what links other forces together unless they act as probes and facts in our laboratory conflicts (1.3.1).

4.6.7 Once we reduce the reduction of "sciences," we are forced to confess that "knowledge" can exist only at the level of traces—in all the senses of this term.

- We often distinguish between the knowledge of the past and that of the modern world (3.2.5, 3.3.0). This is the Great Divide which prevents us from seeing that all these knowledges have the same motor and the same general form: they are not interested in things in themselves, in following them along *their* paths; they are concerned only with man and the modifications to which he can be forced to submit. To speak as we used to, they are "social, too social." To use an image, we could say that ancient falsehoods and modern truths relate to each other like the two revolutions of a single spiral. To be sure the former is smaller than the latter, but they both fall back on society.

  However, they are different, manifestly different. These differences have nothing to do with the critical rigor with which they are elaborated or the presence of data. The difference lies simply *in their size*. In the past only small collectives were fortified. "Things" were pursued simply to pacify them. This knowledge is now said to be false because it was too small. With the building of bigger Leviathans it became necessary to pursue more things for longer, to be more exact, more meticulous, and to reach out into the middle of more forces with more laboratories. But the goal remained the same: it was still man who had to be reformed, deformed, transformed, and informed. Yes indeed, this knowledge which we believe to be new is just as anthropo*morphic* as its predecessors. No, it is even more so! As it becomes necessary to conquer larger numbers of people, it is vital to strike more strongly. So we admire the objectivity of the reasons that we have created? But what do we want to be right about in order to strike so strongly and harshly? Does someone who does not want to kill anyone need facts as hard as clubs?

What about going off at a tangent and following things where *they* take us? Who can honestly say that there are now more people who would be interested in drifting along *their* way than there were in the past? To do this means that we are weak, not strong. It means leaving without thought of return. Or if we do come back, it means that we come with empty hands; with no spoils, trophies, collections, articles, or theses. Can we honestly say that we have seen more people behaving in this way?

The idealists were right: we can only know insofar as we draw things to ourselves. They forgot to add that things have to be drawn together to topple us. Cruise missiles orbit around Leviathans and sooner or later fall back to produce spectacular spin-offs. The Copernican revolution was achieved by ignoring all the rest, and what is left is almost everything. We are left with magic—science and sorcery, future wars, and a certain amount of admirable knowledge obtained, in spite of us all, at the crossroads between anthromorphism and objectivity.

I do not say this because I want to sink our only lifeboat. I say it because I want to prevent shipwreck, or if it is already too late, to make it possible to survive the shipwreck.

4.7.1 Since there are only ties of weakness, there are not two ways of learning—one academic, human, rational, or modern, and the other popular, natural, disorganized, or ancient. There is only one way. We always learn in the same way, without short cuts, foresight, or ever leaving the networks that we build. We make each mistake as many times as is necessary to move from one point to another.

We will never do any better (1.2.1). We will never be able to go any faster. We will never see any more clearly.

- The sciences have always been criticized in the name of superior forms of knowledge that are more intuitive, immediate, human, global, warm, cultivated, political, natural, popular, older, mythical, instinctive, spiritual, or cunning. We have always wanted to criticize science by claiming that an alternative is superior, by adding a court of appeal to the court of first instance, by asking God or the gods to puncture the pride of the learned and to reserve the secret of things for the humble and lowly. But there is no knowledge superior to that of the sciences because there is no scale of knowledge and, in the end, no knowledge at all. We should dissolve all the debates about "degrees of knowledge" into an inferior form of knowledge, the only form that we have. Not metaphysics, but infraphysics. As I have said, we will never be able to rise above unruly politicking (3.6.0).

4.7.2 There is no such thing as superior knowledge and inferior knowledge. If we want to save these terms at all, we will have to say that some forms of knowledge are "higher" than others because the superior have raised themselves with the connivance of the inferior (4.4.0).

4.7.3 Are the "sciences" cold? Rigorous? Inhumane? Objective? Boring? Apolitical? Modern? These unattainable qualities have simply been attributed to them by their enemies who thereby hoped to stigmatize them (Interlude VI). Hot? Disorderly? Violent? Anthropomorphic? Anthropocentric? Interested? Wild? Mythical? No, these do not describe them either. Sparse and fragile, and above all sparse. Their particular sign? No distinguishing marks.

- I do not reproach those poorly conceived aggregates that we call "the sciences" for being too rational, but rather for not understanding the nature of their natures. Let us reduce them to the dimensions that they occupy and finally escape from magic. Since the beginning epistemology has followed in the wake of the sciences, trying to be: PERI-, META-, PARA-, INFRA-, SUPRA-scientific. But this is beating around the bush. Politics is certainly still the best model for trials of weakness, and never more so than when we discover the researcher acting as spokesman for silent crowds of atoms, microbes, or stars. Then we see an executive, a legislative, and a judiciary that for too long have eluded even the most elementary forms of democracy.

4.7.3.1 Those of us who wish to escort "the sciences" back to their proper habitat are more rationalist than most of the learned who

want to extend them "en double." At least we know the cost of the
work involved in multiplying those habitats.

- The gnostics should not misunderstand: I am not trying to make
  their lives easier.

4.7.4 As soon as there is no other world, perfection resides in this
one. Complete knowledge is found in this world as soon as there are
no more degrees of knowledge. The same people who establish degrees
of knowledge are those who then despair of ever reaching the top:
the same reductionists who are alternately drunk with power and
crippled by impotence, arrogant and modest in turn. The trials of
strength are all whole and complete, exact precisely to the extent that
this is possible. *They are not approximate.* Neither are they vague,
conventional, or subjective. Unless new relations of strength are established, they do not have too much or too little. Far from losing
certainty, we finally discover what it was that led to the illusion of a
knowledge beyond uncertainty.

4.7.5 Since there are not two ways of knowing but only one, there
are not, on the one hand, those who bow to the force of an argument,
and, on the other, those who understand only violence. Demonstrations are always of force (3.1.8), and the lines of force are always a
measure of reality, its only measure (1.1.4). We never bow to reason,
but rather to force.

4.7.6 By believing the opposite, we allow certain lines of force and
certain arguments to rule above the networks to which they properly
belong. We create potency (1.5.1), and by so doing, we weaken all
the others.

- There are, they say, reasonable men, who yield only to the force
  of argument, and the remainder, who are unreasonable and submit
  blindly to force without understanding. I have never met anyone
  who did not scorn the unreasonable and who did not believe that
  this scorn epitomized virtue.

4.7.7 As soon as "right" is divided from "might," or "reason" from
"force," right and reason are weakened because we no longer understand their weaknesses, and we steal the only way of becoming
just and reasonable that is available to those who are scorned. These

two losses leave the field free for the wicked. I call this a crime, the only one that we will need in this essay.

- The man who yields to the solidity of a tiny argument only after hundreds of trials and tests, errors and tinkering, in his laboratory nonetheless claims that others who are tested and tried understand nothing and think like morons. Even though he is incapable of clear speech, the moment he steps out of his laboratory door he is outraged to discover "that such a simple argument is not understood by everyone." His outrage nourishes his scorn. Since he despises the fools beneath him, he forgets about the only thing that leads him to yield to the force of this argument: his laboratory, the place where he has been subjected to trials himself. It is a vicious circle. The more foolish the others are, the more he believes that he can "think" and the less he is able to see how he has learned. And the more he extends the potency of reason beyond force, the more reason is weakened.

4.7.8   To oppose right and might is criminal because it leaves the field free for the wicked while pretending to defend it with the potency of what is right. But what is right is without force except "in principle." And so being unable to ensure that what is right is strong, people have acted as though what was strong was wicked. The strong have simply occupied the space left vacant by those who despise them in all innocence.

- As a result of a comprehensible reversal, Machiavelli and Spinoza have been held to be immoral, even though they were right to refuse to distinguish might from right. But the present précis differs from Spinoza's *Tractatus Theologico-Politicus*. Times have changed. The exegesis of religious texts has now been replaced by the exegesis of "scientific" inscriptions. For this reason I think of this essay as a *Tractatus Scientifico-Politicus*. Even so, the object is the same. We are still right at the beginning of the exegesis, and the link between science and democracy has become tenuous in the course of the "wars of science." Like Spinoza, we look cruel in order to be fair.

4.7.9   We do not suffer from a lack of soul, reason, science, or justice, but from a surfeit of all these supplements which are added to relations of force to gear down potency and make the weak impotent. If the weak had in front of them, only the array of weaknesses that I have described, they would dirty their hands and transform it as they pleased.

- "Noli me tangere"—these are the words of a magician who wishes to be both dead and alive at the same time, here and there, strong and rational, strong and right, strong and good.

4.7.10 Since there is only one way of knowing, not two—the testing of relations between forces—there is no way we can avoid a single mistake, absurdity, or crime. We cannot avoid a single experiment or take a single short cut. Even to *think* the contrary is to delude ourselves with criminal illusions.

- How many atomic wars will we have to fight before we yield to the force of the argument that this is no way to conduct our affairs? Listen, it is very simple. We will never do better than those who have simply to convince themselves about trifling matters, have everything they need to hand, and are properly fed, well lit, and appropriately taught. How many mistakes do they make before they start to give up the tiniest prejudice? Tens, hundreds, thousands? So how many wars will it take to convince five billion men and women? Ten? A hundred? Unless, that is, the multitudes can think more quickly and clearly than those in the laboratory.

4.7.11 Those who think that they can do better and work more quickly will always do worse because they will forget to share their only means of knowing and testing. They will believe that they have done enough when they have "diffused" reasons, codes, and results. In fact, all of these wither once they are removed from the scorned networks that keep them strong.

- When Voltaire wanted to pillory religion, he used to sign his letters "écrelinf"—"eradicate the infamous." Religion had done its worst, and more than the worst. Today we find ourselves in the same position. We would never have been able to dream up such a source of marvels, enthusiasm, and warmth, an epiphany to match what we vulgarly call "the sciences." And yet until the millennium ends, we must sign our letters with the same word, "écrelinf." To have knowledge in the next millennium, to be able to talk of exactness without being abused by the irradiated, we must save the knowledge from "the sciences" just as the divine has been saved from the empty shell of religion. Through love of the divine we have had to extirpate everything that was religious within us. Through love of knowledge we must disentangle ourselves from "the sciences."

We cannot balance Galileo against cruise missiles in the way in which the Sermon on the Mount was for so long contrasted with the Inquisition. Apologetics do not interest me. In "science," as in "religion," there are more than enough protestants, mystics, integrists, anabaptists, fundamentalists, and worldy Jesuits. None of them interest me because they all want to reform or regenerate those badly conceived unities, "the sciences." They all seek to reconcile the irreconcilable and, by doing so, make the only thing that I want to understand incomprehensible to me. If cruise missiles gather me in the vineyard, I do not wish to have to bow down before "reason," "erring physics," "the folly of men," "the cruelty of God," or "Realpolitik." I do not wish to invoke muddled explanations which talk of potency when the reason for my death lies in the force of facts. In the few seconds that divide illumination from irradiation I want to be as agnostic as it is possible to be for a man who is present at the passing of the first Enlightenment, as agnostic as it is possible to be for one who is sufficiently sure of both the divine and of knowledge that he dares to hope for the birth of a new Enlightenment. I will not yield to them; I will not believe in "the sciences" beforehand; and neither, afterwards, will I despair of knowledge when one of the relationships of force to which the laboratories have contributed explodes above France. Neither belief, nor despair. I will be as agnostic and as fair as it is possible to be.

So you were wrong, Crusoe. There is no modern world to be set against your primitive island. There is no rational thought to be contrasted with that of the primitive. There are no cultures to be kept apart from the untamed species lurking in the jungle.

We know what happened to Friday and Crusoe when a sailing ship anchored off their island. Tournier has told us (1967/1972). It was Crusoe who remained behind, and Friday who departed. But the following morning Crusoe realized that he was not alone: a ship's boy was there to keep him company.

In our old Europe are we still capable of swapping places in this way?

Bibliography

Notes

Figures

Index

Bibliography

Ackerknecht, Erwin. 1945. "Hygiene in France, 1815–1848." *Bulletin of the History of Medicine* 22:117–155.
—— 1967. *Medicine at the Paris Hospital, 1794–1848.* Baltimore: Johns Hopkins University Press.
Algave, Emile. 1872. "Les réunions scientifiques à l'Assemblée." *Revue Scientifique* 3.2:741–742.
—— 1872. "Editorial." *Revue Scientifique* 3.2:102.
Alix, E. 1881. "Le rôle du médecin dans l'armée. *Revue Scientifique* 11.6:761–764.
—— 1882. "Un mot sur le service sanitaire de l'armée. *Revue Scientifique* 4.2:149–152
Anonymous. 1872. "Editorial." *Revue Scientifique* 3.2:102.
—— 1876. "Congrès international d'hygiène et de sauvetage, 3ᵉ partie." *Revue Scientifique* 22.6:400–410.
—— 1880. "Editorial." *Concours Médical* 1.5:177.
—— 1881. "Le casernement militaire et la société de médecine publique et d'hygiène professionnelle." *Revue Scientifique* 16.7:72–78.
—— 1881. "Editorial." *Revue Scientifique* 30.7:129
—— 1881. "Revue d'hygiène IIᵉ." *Revue Scientifique* 19.3:372–377.
—— 1882. "Editorial." *Revue Scientifique* 23.12:801.
—— 1883. "Revue d'hygiène II." *Revue Scientifique* 24.2:245–250.
—— 1884. "L'exposition d'hygiène de Londres." *Revue Scientifique* 27.9:385–397.
—— 1887. "Editorial" *Concours Médical* 8.10:490.
—— 1887. "Editorial." *Concours Médical* 30.7:362.
—— 1887. "L'oeuvre de la tuberculose." *Revue Scientifique* 2.4:444.
—— 1888. "Editorial." *Concours Médical* 24.11:530

—— 1889. "Causerie bibliographique." *Revue Scientifique* 18.5:628–631.

—— 1893. "A propos de l'origine du typhus: contagion et spontanéité." *Revue Scientifique* 29.4:539–540.

—— 1893. "La variole en Angleterre." *Revue Scientifique* 3.6:699–700

—— 1894. "Editorial." *Concours Médical* 27.10:510.

—— 1894. "La sérothérapie de la diphtérie." *Concours Médical* 15.9:434–

—— 1894. "Variété sur le secret médical." *Concours Médical* 28.4:212.

—— (probably Dr. Jeanne). 1895 "Editorial." *Concours Médical* 13.4:160.

—— 1895. "Lettre à la rédaction." *Concours Médical* 3.8:383.

—— 1900. "Editorial." *Concours Médical* 3.3:97.

—— 1900. "Prix pour le meilleur mémoire sur l'encombrement médical." *Concours Médical* 17.2:79.

—— 1910. "Prophylaxie de la fièvre typhoïde: rapport à l'Académie de Médecine." *Revue Scientifique* 9.4:471–472.

Arloing, S. 1910. "Le présent et l'avenir de la prophylaxie et de la guérison de la tuberculose." *Revue Scientifique* 16.4:481.

Armaingaud. 1893. "La tuberculose." *Revue Scientifique* 14.1:33–42.

Ashmore, Malcolm. 1985. "A Question of Reflexivity: Wrighting (sic). Sociology of Scientific Knowledge" Ph.D. dissertation, University of York.

Augé, Marc. 1975. *Théorie des pouvoirs et idéologie*. Paris: Hermann.

Barnes, Barry. 1974. *Scientific Knowledge and Sociological Theory*. London: Routledge.

Bastide, Françoise. 1985. "Introduction to the Semiotics of Scientific Texts." Mimeographed. Paris: CSI.

Bloor, David. 1976. *Knowledge and Social Imagery*. London: Routledge.

Bouchard. 1895. "Les théories de l'immunité: sérothérapie et vaccination." *Revue Scientifique* 24.8:225–230.

Bouchardat, A. 1873. "Hygiène des hôpitaux: lᵉ partie, l'encombrement nosocomial." *Revue Scientifique* 13:12:552–564.

—— 1879. "Les pestes de Russie." *Revue Scientifique* 29.3:918–922.

—— 1881. "Des principaux modes d'atténuation des microbes ou ferments morbides des maladies contagieuses." *Revue Scientifique* 8.10:458–463.

—— 1883. "Les cinq épidémies de choléra à Paris." *Revue Scientifique* 11.8:170–178.

Bouley, H. 1881. "La nouvelle vaccination." *Revue Scientifique* 29.10:546–550.

—— 1883. "Les découvertes de M. Pasteur devant la médecine." *Revue Scientifique* 7.4:439–443.

Brannigan, Augustine. 1981. *The Social Basis of Scientific Discoveries*. Cambridge: Cambridge University Press.

Braudel, Fernand. 1985. *The Perspective of the World, Fifteenth to Eighteenth Century*. New York: Harper and Row.

Brault. 1908. "Paludisme et maladies parapaludéennes." *Revue Scientifique* 28.3:394–402.

Bullock, W. 1938/1977. *The History of Bacteriology*. New York: Dover.

Callon, Michel. 1980. "Struggles and Negotiations to Decide What Is Problematic and What Is Not: The Sociology of Translation." In *The Social Process of Scientific Investigation*, ed. Karin Knorr, Roger Krohn, and Richard Whitley, pp. 197–220. Dordrecht, Holland: Reidel.

—— 1986. "Some Elements of a Sociology of Translation: Domestication of the Scallops and of the Fishermen of St. Brieuc Bay." In *Power, Action, and Belief: A New Sociology of Knowledge?* ed. John Law, pp. 196–229. Sociological Review Monograph. Keele: Methuen.

Callon, Michel, and Latour, Bruno. 1981. "Unscrewing the Big Leviathans: How Do Actors Macrostructure Reality and How Sociologists Help Them." In *Advances in Social Theory and Methodology: Toward an Integration of Micro and Macro Sociologies,* ed. Karin Knorr and Aron Cicourel, pp. 277–303. London: Routledge.

Callon, M.; Law, J.; and Rip, A., ed. 1986. *Mapping the Dynamics of Science and Technology.* London: Macmillan.

Calmette, A. 1905. "Le rôle des sciences médicales dans la colonisation." *Revue Scientifique* 8.4:417–421.

——— 1912. "Les missions scientifiques de l'Institut Pasteur et l'expansion coloniale de la France." *Revue Scientifique* 3.2:129–133.

Canguilhem, Georges. 1977. *Idéologie et rationalité dans les sciences de la vie.* Paris; Vrin.

Capitan, L. 1894. "Le rôle des microbes dans la société." *Revue Scientifique* 10.3:289–294.

Cartwright, F. 1972. *Disease in History.* London: Rupert Hart Davis.

Carvais, Robert. 1986. "Le microbe et la responsabilité médicale." In *Pasteur et la révolution pastorienne,* ed. Claire Salomon-Bayet, pp. 279–330. Paris: Payot.

Chauffard. 1915. "La guerre et la santé de la race." *Revue Scientifique* 16.23.1.18–25.

Chauveau, A. 1871. "Physiologie générale des virus et des maladies virulentes." *Revue Scientifique* 14.10:362–375.

Chevalier, Louis. 1973. *Laboring Classes and Dangerous Classes During the First Half of the Nineteenth Century in France.* New York: Fertig.

Coleman, William. 1982. *Death Is a Social Disease: Public Health and Political Economy in Early Industrial France.* Madison, Wis.: University of Wisconsin Press.

——— 1985. "The Cognitive Basis of a Discipline: Claude Bernard on Physiology." *Isis* 76:49–70.

Colin, L. 1882. "La fièvre typhoïde dans la période triennale 1877–1878–1879. *Revue Scientifique* 1.4:397–406.

Collins, Harry. 1985. *Changing Order: Replication and Induction in Scientific Practice.* London: Sage.

Corbin, Alain, 1986. *The Foul and the Fragrant,* Cambridge: Harvard University Press.

Corlieu, A. 1881. "L'hygiène à la faculté de médecine de Paris." *Revue Scientifique* 22.10:533–536.

Coutouzis, Mickès. 1983. *Sociétés et techniques en voie de déplacement: le transfert d'un village solaire des Etats Unis en Crête.* Paris: Université Dauphine.

Crosland, Maurice. 1976. "Science and the Franco-Prussian War." *Social Studies of Science* 6:185–214.

Dagognet, François. 1967. *Méthodes et doctrines dans l'oeuvre de Pasteur.* Paris: P.U.F.

——— 1973. *Tableaux et langages de la chimie.* Paris: Le Seuil.

——— 1984. *Ecriture et iconographie.* Paris: Vrin.

242 Bibliography

—— 1984. *Philosophe de l'image*. Paris: Vrin.

Darmon, Pierre. 1982. "L'odyssée pionnière des premières vaccinations françaises au 19ᵉ siècle." *Histoire Economie Société* 1:105–144.

Decaisne, E. 1875. "La protection des nourissons." *Revue Scientifique* 3.4:933–942.

Delaunay, Albert. 1962. *L'Institut Pasteur des origines à aujourd'hui*. Paris: France-Empire.

Deleuze, Gilles. 1968. *Différence et répétition*. Paris: P.U.F.

De Mey, Marc. "Could Cognitive Science Be a Vaccine against Relativistic Sociology of Science?" Paper presented at International Meeting on Scientific Professionalism, Rome, March 20–23, 1985. Mimeographed.

De Raymond, Jean-François. 1982. *Querelle de l'inoculation ou préhistoire de la vaccination*. Paris: Vrin.

Dozon, Jean-Pierre. 1985. "Quand les pastoriens traquaient la maladie du sommeil." *Sciences Sociales et Santé* 3.28–56.

Dubos, Louis. 1961. *The Dreams of Reason: Science and Utopias* New York: Columbia University Press.

Dubos, René. 1951. *Louis Pasteur: Free Lance of Science*. London: Gollancz.

Dubos, René, and Dubos, Jean. 1950. *The White Plague: Tuberculosis, Man, and Society*. London: V. Gollancz.

Duclaux, E. 1879. "Charbon, septicémie, et infection purulente." *Revue Scientifique* 4.1:629–635.

Duclaux, Emile. 1896/1920. *Pasteur: The History of a Mind*, trans. E. F. Smith and F. Hedge. Philadelphia: W. B. Saunders.

—— 1898. *Traité de microbiologie*. Paris: Masson.

Duffy, J. 1979. "The American Medical Profession and Public Health, from Support to Ambivalence." *Bulletin of the History of Medicine* 53:1–23.

Dupuy, Gabriel and Knaebel, G. 1979. *Choix techniques et assainissement urbain en France de 1800 à 1887*. Paris: Institut d'Urbanisme de Paris.

Eisenstein, Elizabeth. 1979. *The Printing Press as an Agent of Change*. Cambridge: Cambridge University Press.

Farley, John. 1978. "The Social, Political, and Religious Background to the Work of Louis Pasteur." *Annual Review of Microbiology* 21:332–342.

Farley, J., and Geison, G. 1974. "Science, Politics, and Spontaneous Generation in Nineteenth-Century France:The Pasteur-Pouchet Debate." *Bulletin of the History of Medicine* 20:257–270.

Favret-Saada, Jeanne. 1977/1980. *Deadly Words: Witchcraft in the Bocage*, trans. C. Cullen. Cambridge: Cambridge University Press.

Fleck, Ludwik. 1935/1979. *Genesis and Development of a Scientific Fact*. Chicago: The University of Chicago Press.

Foucault, Michel. 1963/1973. *The Birth of Clinic: An Archaeology of Medical Perception*. New York: Pantheon.

Foucault, M.; Barret-Kriegel, Thalamy A.; and Beguin, F. 1979. *Les machines à guérir, aux origines de l'hôpital moderne*. Bruxelles: Pierre Mardaga.

Fox, Robert, and Weisz, George. 1980. *The Organization of Science and Technology in France, 1808–1914*. Cambridge: Cambridge University Press

Frazer, W. M. 1950. *A History of Public Health*. London: Baillère, Tindal, and Cox.

Freidson, E. 1970. *The Profession of Medicine: A Study of the Sociology of Applied Knowledge*. New York: Harper and Row.

Fuchs. 1884. "La prophylaxie de l'ophtalmie des nouveaux-nés." *Revue Scientifique* 19.4:493–496.

Furet, François. 1978/1981. *Interpreting the French Revolution*, trans. E. Foster. Cambridge: Cambridge University Press.

Gassot, Dr. 1900. "Lettre à la rédaction sur l'encombrement médical." *Concours Médical* 16.6:284.

Geison, Gerald. 1974. "Pasteur" entry in *The Dictionary of Scientific Biography.* New York: Scribner and Sons, pp. 351–415.

Geison, Gerald, ed. 1984. *Professions and the French State, 1700–1900.* Philadelphia: University of Pennsylvania Press.

Gibert. 1884. "Le choléra à Yport." *Revue Scientifique* 29.11:724–726.

Gibier, Paul. 1893. "Les microbes et la question sociale." *Revue Scientifique* 2.12:722–723.

Goldstein, Jan. 1984. "Moral Contagion: A Professional Ideology of Medicine and Psychiatry in Eighteenth- and Nineteenth-Century France." In *Professions and the French State, 1700–1900*, ed. G. Geison, pp. 181–222. Philadelphia: University of Pennsylvania Press.

Goody, Jack. 1977. *The Domestication of the Savage Mind.* Cambridge: Cambridge University Press.

Gosselin, Dr. 1879. "Editorial." *Concours Médical* 4.10:159.

Goubert, Jean-Pierre. 1985. "L'eau et l'expertise sanitaire dans la France du 19e: le rôle de l'Académie de Médecine et des Congrès Internationaux d'Hygiène." *Sciences Sociales et Santé* 3:75–102.

—— 1986. *La conquête de l'eau.* Paris: Laffont.

Goubert, Jean-Pierre, ed. 1982. "La médicalisation de la société française, 1770–1830." *Historical Reflexions* 9:1–304.

Gouffier, Dr. 1900. "L'encombrement de la profession médicale: causes, résultats, remèdes." *Concours Médical* 10.11:528–554.

Greimas, A. J., and Courtès, J. 1979/1983. *Semiotics and Language: Analytical Dictionary*, trans. L. Cris et al. Bloomington: Indiana University Press.

Grellet, Isabelle, and Kruse, Caroline. 1983. *Histoires de la tuberculose, les fièvres de l'âme, 1800–1940.* Paris: Ramsay.

Guillerme, André. 1983. *Les temps de l'eau, la cité, l'eau, et les techniques: le nord de la France 3e–19e siècle.* Le Creusot: Champvallon.

Haines, Barbara. 1978. "The Interrelations between Social, Biological, and Medical Thought." *The British Journal of History of Science* 11:19–35.

Hannaway, Caroline. 1974. "Medicine, Public Welfare, and the State in Eighteenth Century: The 'Société Royale de Médecine,' 1770–1793." Ph.D. dissertation, Johns Hopkins University.

—— 1972. "The 'Société Royale de Médecine' and Epidemics in the Ancien Régime." *Bulletin of the History of Medicine* 46:259–261.

Hart, Ernest. 1893. "Le berceau du choléra." *Revue Scientifique* 7.10:467–471.

Héricourt, Jean. 1885. "L'influence des milieux sur les microbes pathogènes." *Revue Scientifique* 24.11:525–532.

—— 1888. "Le projet d'organisation de l'hygiène publique." *Revue Scientifique* 25.2:244–249.

Hervouest, Dr. 1894. "Lettre à la rédaction. *Concours Médical* 20.1:26

Herzlich, Claudine. 1982. "The Evolution of Relations between French Physicians and the State from 1880 to 1980." *Sociology of Health and Illness* 4:241–253.

Hesse, Mary. 1974. *The Structure of Scientific Inference.* London: Macmillan.

Holmes, Frederic Lawrence. 1974. *Claude Bernard and Animal Chemistry: The Emergence of a Scientist.* Cambridge: Harvard University Press.

Hughes, Thomas. 1983. "The Electrification of America: The System Builders." *Technology and Culture* 20:124–162.

———— 1983. *Networks of Power: Electric Supply Systems in the United States, England, and Germany, 1880–1930.* Baltimore: Johns Hopkins University Press.

Huguenin, Dr. 1905. "Le diagnostic par les procédés de laboratoire." *Concours Médical* 27.5:202.

Illich, Ivan. 1981. *Limits to Medicine: Medical Nemesis, the Expropriation of Health.* Harmondsworth: Penguin.

Imbert, Jean, ed. 1982. *Histoire des hôpitaux en France.* Toulouse: Privat.

Isambert, François-André. 1985. "Un 'programme fort' en sociologie de la science?" *Revue Française de Sociologie* 26:485–508.

Jeanne, Dr. 1895. "Les cours de bactériologie." *Concours Médical* 4.5:205.

———— 1895. "Editorial." *Concours Médical* 30.3:144

———— 1895. "La bactériologie et la profession médicale." *Concours Médical* 23.3:133

———— 1900. "Editorial." *Concours Médical* 31.3:145–146.

Jewson, N. D. 1976. "The Disappearance of the Sick-Man from Medical Cosmology, 1770–1870." *Sociology* 10:225–244.

Jousset, de Bellesme. 1876. "De la réforme des services sanitaires." *Revue Scientifique* 22.4:401–406.

Jousset, de Bellesme, and Richet, Ch. 1882. "Polémique avec la rédaction." *Revue Scientifique* 22.4:509–510.

Kawabata, Y. 1972. *The Master of Go.* New York: Alfred Knopf.

Keel, Othmar. 1984. "La place et la fonction des modèles étrangers dans la constitution de la problématique hospitalière de l'Ecole de Paris." *History and Philosophy of the Life Sciences* 6:41–73.

Kidder, Tracy. 1981. *The Soul of a New Machine.* London: Allen Lane.

Kirmisson. 1888. "Des réformes urgentes à introduire dans les services de chirurgie." *Revue Scientifique* 10.3:295–301.

———— 1981. *The Manufacture of Knowledge: An Essay on the Constructivist and Contextual Nature of Science* Oxford: Pergamon Press.

Knorr, Karin. 1985. "Germ Warfare." *Social Studies of Science* 15:577–588.

Knorr, Karin; Krohn, Roger; and Whitley, Richard, eds. 1981. *The Social Process of Scientific Investigation.* Dordrecht, Holland: Reidel.

Knorr, Karin, and Mulkay, Michael, eds. 1983. *Science Observed: Perspectives on the Social Study of Science.* Los Angeles: Sage.

Koch, R. 1883. "La vaccination charbonneuse." *Revue Scientifique* 20.1:64–74.

Kottler, Dorian B. 1978. "Louis Pasteur and Molecular Dissymmetry, 1844–1857." *Studies in History of Biology* 2:57–98.

La Berge, Ann F. 1974. "Public Health in France and the French Public Health Movement, 1815–1848." Ph.D. dissertation, University of Tennessee.

———— 1984. "The Early Nineteenth Century French Public Health Movement: The Disciplinary Development and Institutionalization of 'Hygiène Publique.'" *Bulletin of the History of Medicine* 58:363–379.

Lampedusa, Giuseppe D. 1960. *Leopard,* trans. A. Colquhoun. New York: Pantheon.

Landouzy, L. 1885. "L'hygiène à la Faculté de médecine de Paris." *Revue Scientifique* 25.7:97–107.

—— 1909. "L'évolution et le rôle social de la médecine au temps présent." *Revue Scientifique* 7.8:161–170.

Lasalle, Dr. 1888. "Toast au congrés annuel." *Concours Médical* 24.11:562.

Latour, Bruno. 1981. "Who Is Agnostic? Or What Could It Mean to Study Science?" In *Knowledge and Society: Studies in the Sociology of Culture Past and Present*, ed. R. Jones and R. Kuclick. Greenwich: JAI Press, 3:199–216.

—— 1983a. "Comment redistribuer le Grand Partage?" *Revue de Synthèse* 110:202–236.

—— 1983b. "Give Me a Laboratory and I Will Raise the World." In *Science Observed: Perspectives on the Social Study of Science*, ed. K. Knorr and M. Mulkay. Los Angeles: Sage

—— 1986a. "The Powers of Association." In *Power, Action, and Belief: A New Sociology of Knowledge?* ed. John Law. Sociological Review Monograph. Keele: Methuen, pp. 264–280.

—— 1986b. "Visualization and Cognition." In *Knowledge and Society: Studies in the Sociology of Culture Past and Present*, ed. H. Kuclick. Greenwich: JAI Press, 6:1–40.

—— 1987. *Science in Action: How to Follow Scientists and Engineers Through Society*. Cambridge: Harvard University Press.

—— 1988. "A Politics of Explanation." In *Knowledge and Reflexivity*, ed. S. Woolgar. London: Sage.

Latour, Bruno, and Bastide, Françoise. 1986. "Fact and Fiction Writing." In *Mapping the Dynamics of Science and Technology*, ed. M. Callon, J. Law, and A. Rip. London: Macmillan.

Latour, Bruno and de Noblet, Jocelyn, eds. 1985. *Les vues de l'esprit: visualisation et connaissance scientifique*. Special issue of *Culture technique*.

Latour, Bruno, and Fabbri, Paolo. 1977. "Pouvoir et devoir dans un article de science exacte." *Actes de la Recherche en Sciences Sociales* 13:81–99.

Latour, Bruno, and Woolgar, Steve. 1979/1986. *Laboratory Life: The Construction of Scientific Facts*. Los Angeles/Princeton: Sage/Princeton University Press.

Law, John. 1986. "On the Methods of Long-Distance Control, Vessels, Navigation, and the Portuguese Route to India." In *Power, Action, and Belief: A New Sociology of Knowledge?* ed. John Law. Sociological Review Monograph. Keele: Methuen, pp. 234–263.

—— 1986. *Power, Action, and Belief: A New Sociology of Knowledge?* Sociological Review Monograph. Keele: Methuen.

Lécuyer, Bernard-Pierre. 1977. "Démographie, statistique, et hygiène publique sous la monarchie censitaire." *Annales de Démographie Historique* 3:215–248.

—— 1986. "L'hygiène en France avant Pasteur." In *Pasteur et la Révolution Pastorienne*, Paris: Payot. ed. Salomon-Bayet, pp. 65–139.

Leduc, Stéphane. 1892. "Les conditions sanitaires en France." *Revue Scientifique* 20.2:232–239.

Lemaine, G. and MacLeod, R. eds. 1976. *Perspectives on the Emergence of Scientific Disciplines*. The Hague/Paris: Mouton.

Lemure, Jean. 1896. "Morbidité et mortalité pendant l'expédition de Madagascar." *Revue Scientifique* 11.1:47–51.

Léonard, Jacques. 1967. *Les officiers de santé de la Marine française de 1814 à 1835.* Paris: Klinsieck.

——— 1979. *Les médecins de l'Ouest au 19ᵉ siècle.* Lille: Atelier de Reproduction de l'Université de Lille.

——— 1981. *La médecine entre les pouvoirs et les savoirs: histoire intellectuelle et politique de la médecine française au 19ᵉ siècle.* Paris: Aubier.

——— 1986. "Comment peut-on être pasteurien?" In *Pasteur et la révolution pastorienne,* ed. Claire Salomon-Bayet, pp. 143–179. Paris: Payot.

Loye, P. 1885. "Les microbes bienfaisants." *Revue Scientifique* 14.2:214–216.

Martin. 1880. "Les revendications de l'hygiène publique en France." *Revue Scientifique* 8.5:1062–1071.

McNeill, William. 1976. *Plagues and Peoples.* New York: Anchor Press.

——— 1982. *The Pursuit of Power: Technology, Armed Forces, and Society since A.D.1000.* Chicago: University of Chicago Press.

Merton, R. K. 1973. *The Sociology of Science; Theoretical and Empirical Investigations* Chicago: University of Chicago Press.

Mollaret, Henri, and Brossollet, Jacqueline. 1984. *Alexandre Yersin ou le vainqueur de la peste.* Paris: Fayard.

Murard, Lion, and Zylberman, Patrick. 1984. "De l'hygiène comme introduction à la politique expérimentale, 1875–1925." *Revue de Synthèse* 105:313–342.

——— 1985. "La raison de l'expert ou l'hygiène comme science sociale appliquée." *Archives Européennes de Sociologie* 26:58–89.

Murphy, Terence. 1979. "The French Medical Profession's Perception of Its Social Function between 1776 and 1830." *Medical History* 23:259–278.

——— 1981. "Medical Knowledge and Statistical Methods in Early Nineteenth-Century France." *Medical History* 25:301–319.

Musil, Robert. *A Man without Qualities.*

Nattan-Larrier, L. 1915. "Les grandes étapes de la protistologie pathologique coloniale." *Revue Scientifique* 10.7:294–303.

Nicol, Louis. 1974. *L'épopée pastorienne et la médecine vétérinaire.* Garches: chez l'auteur.

Nicole, Jacques. 1953. *Un maître de l'enquête scientifique: Louis Pasteur.* Paris: Vieux Colombier.

Nye, Robert. 1984. *Crime, Madness, and Politics in Modern France: The Medical Concept of National Decline.* Princeton: Princeton University Press.

Pasteur, Louis. 1871. "Pourquoi la France n'a pas trouvé d'hommes supérieurs au moment du péril." *Revue Scientifique* 22.7:73–77.

——— 1871. *Quelques réflexions sur la science en France.* Paris: Gauthier-Villars.

——— 1883. "La vaccination charbonneuse: réponse au docteur Koch par M. Pasteur." *Revue Scientifique* 20.1:74–84.

——— 1922/1939. *Oeuvres complètes.* Paris: Masson.

Pasteur, Louis, and Thuilier, Louis. 1883. "La vaccination du rouget des porcs à l'aide du virus mortel atténué de cette maladie." *Revue Scientifique* 1.12:673–675.

——— 1980. *Correspondence of Pasteur and Thuilier concerning Anthrax and Swine Fever Vaccinations,* ed. D. Wrontnowska. Alabama: University of Alabama Press.

Péguy, Charles. 1914. *Clio: dialogues de l'âme païenne et de l'âme charnelle*. Paris: Pleïade.

Peter. 1883. "Réponse de M. Peter à M. Pasteur." *Revue Scientifique* 5.5:557–561.

Pickstone, J. V. 1981. "Bureaucracy, Liberalism, and the Body in Post-Revolutionary France: Bichat's Physiology." *History of Science* 19:115–142.

Pinch, Trevor. 1986. *Confronting Nature: The Sociology of Solar Neutrino Detection*. Dordrecht, Holland: Reidel.

Pluchon, Pierre, ed. 1985. *Histoire des médecins et pharmaciens de marine et des colonies*. Paris: Privat.

Ramsey, Matthew. 1977. "Medical Power and Popular Medicine: Illegal Healers in Nineteenth Century France." *Journal of Social History* 10:560–587.

—— 1984. "The Politics of Professional Monopoly in Nineteenth-Century Medicine: The French Case and Its Rivals." In *Professions and the French State 1700–1900*, ed G. Geison, pp. 225–306. Philadelphia: University of Pennsylvania Press.

Reclus, P. 1890. "Les origines et les tendances de la chirurgie contemporaine." *Revue Scientifique* 25.1:104–112.

Reiser, S.L. 1978. Medicine and The Reign of Technology. Cambridge: Cambridge University Press.

Reynaud, P. 1881. "Editorial." *Concours Médical* 29.1:102.

Richard, E. 1883. "Le parasite de l'impaludisme." *Revue Scientifique* 27.1:113–118.

Richet, Ch. 1880. "Editorial." *Revue Scientifique* 10.7:35.

—— 1881. "Editorial." *Revue Scientifique* 5.2:161

—— 1882. "Editorial." *Revue Scientifique* 15.4:449

—— 1883. "Editorial." *Revue Scientifique* 24.2:225

—— 1886. "Editorial." *Revue Scientifique* 6.3:289.

—— 1888. "La physiologie et la médecine." *Revue Scientifique* 24.3:352–362.

—— 1889. "Commentaire de la rédaction." *Revue Scientifique* 16.3:330.

—— 1889. "L'hygiène et la mortalité à Paris." *Revue Scientifique* 16.11:636–638.

—— 1894. "La prophylaxie de la diphtérie." *Revue Scientifique* 30.9:412–413.

—— 1895. "La sérothérapie et la mortalité de la diphtérie." *Revue Scientifique* 20.7:65–69.

Robinet, G. 1881. "Les prétendus dangers des cimetières." *Revue Scientifique* 18.6:779–782.

Rochard, 1887. "L'avenir de l'hygiène." *Revue Scientifique* 24.9:387–395.

Roll-Hansen, Nils. 1979. "Experimental Method and Spontaneous Generation: The Controversy between Pasteur and Pouchet, 1859–1864." *Journal of the History of Medicine* 34:273–292.

Roux, E. 1915. "Jubilé de Metchnikoff. *Annales de l'Institut Pasteur*, August, pp. 410–230.

Rozenkranz, Barbara. 1972. *Public Health in the State: Changing Views in Massachusetts, 1843–1936*. Cambridge: Harvard University Press.

Rudwick, Martin S. 1985. *The Great Devonian Controversy: The Shaping of*

Scientific Knowledge among Gentlemanly Specialists. Chicago: University of Chicago Press.

Salomon-Bayet, Claire. 1978. *L'institutionalisation de la science et l'expérience du vivant: méthodes et expérience à L'Académie Royale de Médecine*. Paris: Flammarion.

Salomon-Bayet, Claire, ed. 1986. *Pasteur et la révolution pastorienne*. Paris: Payot.

Serres, Michel. 1980. *Le Passage du Nord-Ouest (Hermès V)*. Paris: Minuit.

——— 1980/1982. *The Parasite*, trans. L. R. Schehr. Baltimore: John Hopkins University Press.

——— 1983. *Hermes: Literature, Science, Philosophy*, trans. J. Harari and D. E Bell. Baltimore: Johns Hopkins University Press.

Shapin, Steve. 1982. "History of Science and Its Sociological Reconstruction." *History of Science* 20:157–211.

Shapin, Steve, and Schaffer, Simon. 1985. *Leviathan and the Air-Pump*. Princeton: Princeton University Press.

Shapiro, Ann-Louise. 1980. "Private Rights, Public Interests, and Professional Jurisdiction: The French Public Health Law of 1902." *Bulletin of the History of Medicine* 54:4–22.

Shryock, Richard Harrison. 1936/1979. *The Development of Modern Medicine: An Interpretation of the Social and Scientific Factors Involved*. Madison: University of Wisconsin Press.

Spinoza. 1665. *Theologico-Political Treatise*, trans. R. H. Ewes. New York: Dover.

Starr, Paul. 1982. *The Social Transformation of American Medicine: The Rise of a Sovereign Profession and the Making of a Vast Industry*. New York: Basic Books.

Stepan, N. 1978. "The Interplay between Socio-Economic Factors and the Medical Science: Yellow Fever Research Cuba and the United States." *Social Studies of Science* 8.4:397–429.

Sternberg, G. 1889. "Les bactéries." *Revue Scientifique* 16.3:326–330.

Stokes, W. 1872. "La médecine publique en Angeleterre." *Revue Scientifique* 6.7:13–21.

Sussman, George D. 1977. "The Glut of Doctors in Mid-Nineteenth Century France." *Comparative Studies in Society and History* 19:287–304.

Tolstoy, Leo. 1869/1986. *War and Peace*, trans. R. Edmonds. Harmondsworth: Penguin.

Toulouse, P. 1905. "La réforme des études médicales." *Revue Scientifique* 25.11:702–703.

Tournier, Michel. 1967/1972. *Friday and Robinson: Life on Esperanza Island*, trans. R. Manheim. New York: Knopf.

Trélat, Emile. 1890. "Contribution de l'architecte à la salubrité des maisons des villes." *Revue Scientifique* 7.6:705–711.

——— 1895. "La salubrité." *Revue Scientifique* 10.8:163–170.

Tyndall, J. 1876. "La putréfaction et la contagion dans leurs rapports avec l'état optique de l'atmosphère." *Revue Scientifique* 10.6:560–564.

——— 1877. "La fermentation et ses rapports avec les phénomènes morbides." *Revue Scientifique* 17.2:789–800.

Valentino. 1904. "Critique du secret médical, partie 1ᵉ." *Revue Scientifique* 17.9:353–357.

Valléry-Radot, R. 1911. *La vie de Pasteur* Paris: Hachette.

—— 1956. *Images de la vie et de l'oeuvre de Pasteur: documents photographiques*. Paris: Flammarion.

Watkins, Dorothy E. 1984. "The English Revolution in Social Medicine, 1889–1911." Ph.D. dissertation, University of London.

Weiner, Dora B. 1968. *Raspail: Scientist and Reformer*. New York: Columbia University Press.

Weisz, George. 1978. "The Politics of Medical Professionalization in France, 1845–1848. *Journal of Social History* 12:1–30.

—— 1980. "Reform and Conflict in French Medical Education, 1870–1914." In *The Organisation of Science and Technology in France, 1808–1914,* ed. R. Fox and G. Weisz. Cambridge/Paris: Cambridge University Press/MSH, pp. 61–94.

Worboys, Michael. 1976. "The Emergence of Tropical Medicine: A Study in the Establishment of a Scientific Speciality." In *Perspectives on the Emergence of Scientific Disciplines,* ed. G. Lemaine et al. The Hague/Paris: Mouton.

Yersin, A. 1894. "La peste bubonique à Hong-Kong." *Annales de l'Institut Pasteur,* August, pp. 663–672.

Notes

Introduction. Materials and Methods

1. L. Tolstoy (1869/1986). All references are to the one-volume 1986 Penguin edition.

2. This book owes a great deal to Michel Serres' work, especially to his geographical metaphor of the Northwest Passage (Serres: 1980). Instead of envisaging the divide between human and natural sciences as something simple, like a strait, Serres offers the image of a multiplicity of islands, channels, peninsulas, dead ends, and narrow paths, as confusing and as beautiful as a map of the Northwest Passage.

Human Sciences vs. Natural Sciences

3. See M. Serres (1983) for the association of the military, scientists, and businessmen in a thanatocracy more powerful than all the demo-, techno-, and autocracies of the past.

4. Tolstoy himself proposes a global religious explanation for the vast movement that freed the Russians from Napoleon's army. This explanation revolves around God's providential plan for Russia.

5. Spinoza's treatise (around 1665) comprises a long scientific analaysis of the Old Testament, thus establishing the modern way of doing biblical exegesis. This analysis helped redefine the relations between political power, freedom of conscience, and religious revelation. At the center of the treatise is a superimposition of might on right that Spinoza deemed necessary in order to put democracy on firm ground. Without claiming to emulate Spinoza any more than I tried to imitate Tolstoy, I used both as guides and protectors.

6. If relation of "forces" reminds readers too much of Nietzsche's will to

power, let them replace "forces" by "weaknesses." This replacement is a sure remedy against misunderstanding the main point of this book.

7. According to most philosophies of science, empirical studies by historians, economists, and sociologists are too feeble to make up the whole picture of what science is about. Why? Because case studies, philosophers argue, do not concern themselves with the "foundations" of science or with the "transcendental conditions" of any argumentation. There is thus a division of labor between philosophers of science, who think they have a perfect right to ignore (and even to despise) empirical studies, and the social scientists, who think they should never indulge in philosophical arguments. It is only reluctantly that an "epistemologically relevant sociology of science" was invented, as if the honor of feeding a few case studies to transcendental philosophy was finally granted to scholars in the social sciences. This division of labor is a catastrophe; philosophy and field studies should be carried out under the same roof and, if possible, in the same head. I use philosophy here in the same way that theories are used in the other sciences, that is, to designate, highlight, anticipate, underline, dramatize, tie together the empirical material. I use philosophy not because empirical material *lacks* some "foundation" but because I want, on the contrary, more details, more materials, more historical case studies. In another book (1987) I put forward similar arguments, but at a third level which is intermediary between case studies and philosophy. *Science in Action* is thus a companion book to the present one.

8. See e.g. F. Dagognet (1967, p.212); R. Valléry-Radot (1911).

9. This freedom of choice of the metalinguistic level required for the explanation, taken from the methodological principle common to all the other social sciences, forms the basis of much anthropology of science. See B. Latour and S. Woolgar (1979/1986); B. Latour (1981). But it raises many aporetic consequences, which are nowhere more ironically illustrated than in M. Ashmore (1985). If the epistemological necessity of this freedom does not seem obvious to the reader, it can be defended on stylistic grounds: it leads to a *multiplication* of the languages available to talk about science.

10. According to several reviews of the French edition of this book, I failed pitifully on three grounds. K. Knorr (1985), F.-A. Isambert (1985), M. de Mey (1985), and Salomon-Bayet (1986) praise the work for its social and political interpretation of Pasteur's "manipulation," "exploitation," and "clever opportunism," and for the nice way in which I put aside technical contents and limit myself to the application of science to society! Although no text can defend itself against its readers' interpretations, I want to stress again that I am not interested here in offering a social or a political explanation of Pasteur as an alternative to other cognitive or technical interpretations. I am interested only in retracing our steps back to the moment when the very distinction between content and context had not yet been made. If I use the words "force," "power," "strategy," or "interests," their use has to be equally distributed between Pasteur and those human or nonhuman actors who give him his strength. See M. Callon (1986).

11. I use "actor," "agent," or "actant" without making any assumptions about who they may be and what properties they are endowed with. Much more general than "character" or "dramatis persona," they have the key feature of being autonomous figures. Apart from this, they can be anything—individual ("Peter") or collective ("the crowd"), figurative (anthropomorphic or zoomorphic) or nonfigurative ("fate"). A. Greimas and J. Courtès (1979/1983). See also Part Two.

12. When there is no journal title after a quotation, the *Revue Scientifique* is meant.

13. For an introduction to semiotics as applied to scientific texts, see F. Bastide (1986). I constantly use the notions of performance—what characters do—and competence—what this action implies (see A. Greimas and J. Courtès: 1979)—to define the actors (or actants or agents) that comprise the characters of this narrative.

14. On the Franco-Prussian War and its effect on French science, see M. Crosland (1976).

15. Historians share with sociologists a belief in the existence of a context *in which* the events have to be carefully situated. For sociologists this context is made up of the social forces that explain the events (the catch phrases including "it is no coincidence that" or "it fits in well with the interest of"); for historians the context is a set of events firmly tied to the chronological framework. For both trades there exists a context and it is retrievable, at least in principle. Despite their feud, the two disciplines believe in the difference between context and content. Once this belief is shared, people can disagree, some preferring to stick to the content (they are called internalists), others to the context (they are called externalists), and still others to a careful balance between the two. For the two disciplines, additional sources will make the series *converge* into one overall more or less coherent picture. This is the basic assumption that is not shared by semioticians, or for that matter by ethnomethodologists. More data, more sources will make the sources *diverge* more and more. To be sure, it might be possible to obtain some effects of totality, but these are exceptions, local productions inserted among the others and dependent upon a local panopticon. It is because this book relies on semiotics that it is neither history nor sociology. It explores different assumptions about what composes both content and context and different ways of constituting this mixture.

16. This notion of translation has been developed by M. Callon (1986). M. Callon, J. Law, and A. Rip, eds. (1986), and B. Latour (1987) and applied to the study of science and technology in order to fuse the notions of interest and research program in a more subtle way. First, translation means drift, betrayal, ambiguity (1.2.1). It thus means that we are starting from *inequivalence* between interests or language games and that the aim of the translation is to render two propositions equivalent. Second, translation has a strategic meaning. It defines a stronghold established in such a way that, whatever people do and wherever they go, they have to pass through the contender's position and to help him further his own interests. Third, it has a linguistic sense, so that one version of the language game translates all the others, replacing them all with "whatever you wish, this is what you really mean."

17. See F. Bastide (1986); B. Latour and P. Fabbri (1977); B. Latour and F. Bastide (1986).

1. Strong Microbes and Weak Hygienists.

1. L. Tolstoy (1869/1986) in the epilogue to *War and Peace,* criticizes mystical as well as social explanations of strategy. His critique of the notion of power is especially interesting for us (p.1409). There is no gain to be had going from the "internalist" notion that ideas have an internal thrust of their own to the "externalist" notion that people have political power. The notion of power, as

well as of planned strategy, simply disguises our ignorance. B. Latour (1986a). On the difference between force and potency, see Part Two.

2. No distinction is made here between science and technology. The mechanisms that *trans*form what is *trans*ported are the same. On the distinction between the diffusion model and the translation model, see B. Latour (1987, ch. 3).

3. The active society that makes up immense parts of bacteriology is not the same as the society used as a backdrop or a "social context" for the history of science. Herein originates the misunderstanding between microsociologies of science and philosophies of science. Society has to be redefined in order to become usable in "social" studies of science.

4. Among many useful references, see L.Chevalier (1973); A. Corbin (1986); L. Murard and P. Zyberman (1984, 1985); W. Coleman (1982); R. Nye (1984).

5. The fight against *degeneration* (which is not at all a fight against microbes) could have done everything that was accomplished with the hybrid Pasteurism-Hygiene. R. Nye (1984) makes the most thorough study of degeneration: "By the turn of the century, a medical outlook of *bio-pouvoir* had thoroughly penetrated popular consciousness. A medical theory of regeneration was so successful in integrating the palpable and familiar litany of social pathologies into a discourse of national decline that it escaped the terminological prison of the clinic and throve in the arena of public debate" (p.170).

6. The "addresser" communicates to the "addressee" not only the competence but also the values that are at stake in the narration. See A. Greimas and F. Courtès (1979/1983). In this sense the "necessary movement of regeneration" is never discussed because it is what gives everyone the "right" to discuss.

7. See W. F. Frazer (1950).

8. W. Coleman (1982) studies mostly Villermé and his school over the fifty-year period before Pasteur's takeover of French medicine. "Public health investigation was a distinctive feature of 19th century European society. Interest in, broadly speaking, the sanitary conditions of discrete populations easily crossed boundaries and created, within two generations, a recognizable medical speciality. The hygienists were armed with novel conceptual and methodological tools, they soon won academic and other employment, and they were backed by remarkable public interest in their undertakings. Both British and French physicians had given early stimulus to this movement. In the quarter century after the Congress of Vienna, however, leadership passed to France; and it was there, principally in Paris, that *hygiène publique,* or public health, won formal constitution as a science." (p.xvi).

9. This conflict is the drama of Villermé's life and is what renders Coleman (1982) so beautiful. "The hygienists' position was marked by a continuing tension. None knew better than they the nature and probable sources of human suffering in a rapidly urbanizing and industrializing society. But their remedies for these problems almost always stopped short of requiring major social change" (p.22). This contradiction between political economy annd hygiene is what the definition of bacteriology will resolve in part by shifting the interest from "sick paupers" to "dangerous microorganisms." The contradiction will be alleviated because many precautions suggested by the health movement will no longer be necessary within the bacteriological treatment of the same problem.

10. The link between mortality and class created by Villermé is as interesting as the link between attenuated microbes and diseases later created by Pasteur.

They are both defined by "laboratory" methods, except that in Villermé's case the laboratory is Paris checkered with statistical institutions. See Coleman (1982): "Paris was vast, it was diverse, its toll of mankind seemed beyond necessity and justice. The city, through its vital statistics and public practices, was to become a laboratory, a centre for social discovery if not yet social amelioration. The city thus gave the hygienists their great opportunity" (p.43). Villermé's definitions, like Pasteur's, are at variance with interpretations of diseases as due to crowding or to environmental factors alone.

11. The very definition of a context, of economic trends, of an historical "longue durée," are the outcome of a set of social sciences (sociology, economics, history). A dedicated sociologist of science cannot criticize the natural sciences while uncritically believing in the social ones. Consequently, a new principle of *symmetry* has to be defined which requires us to maintain the same critical stand with respect to society and nature. The "social context" can never be used to "explain" a science. See B. Latour (1987. chs. 3, 6).

12. Statistics is the prior science, the one that created epidemics and epizootics as recognizable entities. See E. Ackerknecht (1945); B.-P. Lécuyer (1977); W. Coleman (1982); T. Murphy (1981). On the earlier period, see A. La Berge (1974).

13. In saying this, I am not committing a sin against M. Rudwick's rule that a narrative should never be retrospective (1985). I am, on the contrary, reconstructing the movement of hygiene left to itself, before the advent of Pasteurism. Pasteurian victory has been so complete that it is difficult to recapture the requirements that Pasteurians had to meet in order to be believed at all. This does not mean that Pasteur's interests "fitted" those of the hygienist, but that there was room for a *negotiation* about the meaning of contagion if, and only if, the Pasteurians were able to take into account the variability of the contagion.

14. We should never sever a social movement from the army of journalists, thinkers, social scientists, and politicians that "socially constructs" it. Thus, "social movement" is used here as an abbreviation to designate the work of composition, definition, aggregation, and statistics already done by the hygienists and their troops. I am not using it as a social "cause" that explains the science, but as the reified result of an earlier politicoscientific imbroglio.

15. See W. Coleman (1982): "As noted, hygienists were not uninformed regarding disease theory; their concern simply was directed to other matters, matters that were "biological" in a different and, if the expression be permitted, more expansive sense. The hygienist attended to the essential conditions of existence— food; supply and purity of water; presence and absence of human, animal, and other wastes; the conditions of bodily and mental activity, including above all work, shelter, or protection from the elements—and realized that all of those possessed an underlying economic character; the environment was thereby rendered social in nature. The hygienist also realized that this socioeconomic dimension touched directly upon disease *sensu strictu*" (p.202). Even the link between contagion, social theory, and medical power could have been made without the remotest tie with bacteriology. See J. Goldstein (1982).

16. On the dispute about the general factors that caused the long-term decline in infectious diseases, see R. Dubos (1961); I. Ilich (1981).

17. See e.g. R. H. Shyrock (1936/1979): "The result was that the health program entered a new phase after 1870; so impressive a phase that it was soon viewed as the very beginning of things in public hygiene. There ensued a tendency to give too much credit to leaders in medical research; whereas up to 1870 they

received too little" (p.247). See also W. Bullock (1938/1977); W. Frazer (1950).

18. The ability of a scientific proof to convince has a multiplicity of causes, not any single one. This has been "proven" in several case studies which constitute most of the social studies of science paradigm. See esp. K. Knorr (1981); H. Collins (1985); T. Pinch (1986). To be more reflexive, I should say: believing that evidence of the underdetermination of scientific proofs has been offered by these case studies is a sure sign that we share the same professional commitment.

19. Historians of Pasteurism naturally describe more opponents, many of whom were actually provoked by Pasteur's sometimes abrupt remarks. See e.g. J. Farley and J. Geison (1974) on Pouchet; L. Nicol (1974) and D. Weiner (1968) on Raspail. I should remind the reader again at this point that I am limiting my sources to what an "ideal" reader would know of Pasteur and his alliances, were he or she to read only the *Revue Scientifique*. A little more information on conflicts can be gathered from Salomon-Bayet, ed. (1986).

20. G. Canguilhem (1977). This germ theory of the germ theory was very frequent in Pasteur's time. It has continued to the present as one of the many agricultural metaphors used by historians of science and technology in replacing the *composition* of science by its *unfolding*. It is an avatar of the notion of "potency" studied in Part Two (2.1.3).

21. On Koch's aborted attempt, see R. Dubos (1950). The two words "credulity" and "credibility" share the same beginning and indeed the same root; all that distinguishes them is the outcome of a struggle: the losers were credulous and the winners credible. David Bloor (1976) has most clearly defined the task of any sociology of science by introducing the notion of symmetry. The losers and the winners must be studied in the same way and explained with the same set of notions. If the evolution of our field has made the notion of a "social" explanation obsolete, the principle of symmetry remains the basis of most work in the area.

22. This *addition* never appears enough to those who wish to provide a demiurgic interpretation of science; they want science to generate all its content from within itself, and they regard as dangerous reductionists those who produce it from its context. Yet this same addition appears too much to those who wish to offer a social rendering of science; they would like to explain a science because it *fits well with* other interests, and they consider as internalists those who deny the notion of a fit. I am weaving my way between these two reductionisms. There is nothing to be gained in limiting the cause of the spread of a innovation to any one member of the chain: everyone is defining what society is about, including of course the scientists themselves.

23. The consideration of hygiene as a means of social control is a common thread to much nineteenth-century history. For the development of ideas in France close to those of Foucault around the concept of "biopower," see L. Murard and P. Zylberman (1984, 1985); A Corbin (1982); B.-P Lécuyer (1986). However, I am interested here not in the predictable *application* of a given power on the bodies of the wretched and the poor but in the earlier *composition* of an unpredictable source of power. It is precisely at the time when no one can tell whether he is dealing with a new source of power that the link between science and society is most important. When almost everyone is convinced, then, but only then and afterward, will hygiene be a "power" to discipline and to coerce (see ch. 3).

24. B. Rozenkranz (1972) reconstitutes the *accusation* process and its varia-

tions: who should be blamed for what sort of evil? In this sense her work is very close to that of many anthropologists. Bacteriology reshuffles those who are responsible for the spread of diseases, who are poor and dirty, who are contagious and rich. Speaking of the arrival of scientists on the Board of Health in Boston, she writes: "Their focus on the bacteriological etiology of preventable diseases placed responsibility for negligence firmly in the hands of the powerful rather than the weak. In the process of establishing the vigor and competence of the biological sciences in preventive hygiene, they challenged the identity of filth and disease and refined both the ideology and program of public health" (p.98). Others fight this new definition of the social link: "Reliance on pasteurization would, in Walcott's view, terminate the ultimate responsiblity of the individual to preserve conscientious cleanliness . . . For Walcott, whose concept of prevention rested on enlightened restraint of the individual rather than the bacteriological organism, the price [of pasteurization] was too high to pay" (p.110).

25. A.-L Shapiro (1980) makes a similar argument at the level of political philosophy during the same period: "More and more the concept of 'solidarism' crept into official pronouncements and became the characteristic social philosophy of the Third Republic. It provided the means to steal the thunder from the socialists while justifying a limited, but legitimate, extension of the powers of the State. Solidarism emphasized the inter-dependence of all members of society and used the vocabulary of contractual obligation to demonstrate that each individual was responsible for the well-being of all and must, therefore, be willing to sacrifice some elements of personal liberty in the interest of the community. Public health became a quintessential example of the practical application of solidarism" (p.15).

26. I am fusing here the *method* of semiotics with an *argument* from sociology. My claim is simply that the lists of actors and associations obtained by a semiotic study of the articles of the period are longer and more heterogeneous than the lists offered by the sociologists or social historians of the period. To grasp the argument of the next section, we must accept a certain degree of ignorance as to what is the real list of actors making up a society, and a certain degree of agnosticism about which are human and which are nonhuman, which are endowed with strategy and which are unconscious. Because of this fusion, this ignorance, and this agnosticism I prefer to call the discipline I work in "anthropology of science and technology." When ethnographers work in exotic realms, they often gain, without too much ado, this state of uncertainty—or of grace— that is so hard to get when treating *our* societies. See 3.5.2; B. Latour (1983a).

27. Viewed in this way, the research program of T. Merton (1973) and of most American sociology *of scientists* seems more reasonable. American sociologists, knowing that they did not have a sociology capable of studying the contents of science, limited themselves to its context, to rewards, citations, and careers—that is, to what sociologists knew best how to do. By contrast, the British school courageously entered into the content, despising this American sociology *of scientists* that was doing only half the job. See D. Bloor (1976); H. Collins (1985); S. Shapin (1982). In spite of its great achievements, this enterprise appears disappointing because the contents and the contexts remain very far apart. Most of the sociology of science is internalist epistemology sandwiched between two slices of externalist sociology. We are now at a new crossroads: we must either give up studying the contents of science or change the sociology we started with.

28. Conservatism, Catholicism, love of law and order, fidelity to the Empress,

brashness, passion—those are approximately all we get of the "social factors" acting on Pasteur. R. Dubos (1951); J. Farley and G. Geison (1974); J. Farley (1978). They are enough to provoke the rationalist, who is shocked by such an intrusion of social elements into the pure realm of autonomous science. N. Roll-Hansen (1979). But they are not much if we put on the other side all the scientific work to be explained. This imbalance does not disturb the sociologist, who explains many different things with the same word, believing that these words have some causal potency that enable them to generate many different effects. Nor does the imbalance disturb the social historian, who needs social explanation simply to sketch the background of Pasteur's work and then quickly to return to classic internalist studies. But it does disturb me if I wish to give an irreductionist explanation of the content itself: the explanation has to be at least as rich as the content, not poorer.

29. See M. Serres (1980/1982, 1983). The main importance of his philosophy for the study of science is that he is one of the few philosophers to be utterly uninterested in the notion of a critique, be it transcendental or social. As a consequence, he makes no distinction between language and metalanguage, using a poem, a myth, a theorem, or a machine as something that explains as well as something to be explained.

30. See M. Callon and B. Latour (1981). If we trace in the dictionary the slow drift of *socius* with its associated or successive meanings, we will be struck by how the meaning of "social" has continued to shrink (3.4.7.). It begins as "association" and ends up with "social workers" by way of the "social contract" and the "social question." My redefinition aims simply to resurrect its original richness of meaning.

31. W. McNeill (1976) is the inspiration of these pages. W. McNeill (1982) is most relevant for analysis of the politicoscientific imbroglio.

32. The very distinction between science and society is thus an *artifact* of the attribution process, exactly as the notion of a man's power is, for Tolstoy, an artifact of the historian's description (1869/1986, pp. 1409). On this critique of power, see J. Law, (1986).

33. Is this enough to convince the reader that I am not using an argument in terms of a science "fitting in well with" its context? The whole of hygiene (as well as the whole of bacteriology) is displaced and translated. What makes the reader immediately translate this argument into a reductionist social explanation is the remaining notion of a cause. Hygiene does not cause bacteriology any more than it fits in well with bacteriology. The two associate their common weaknesses and renegotiate the meaning of their alliance. Anyway the notion of a "cause" is one of many avatars of "potency" (2.1.6). A cause is always the *consequence* of a long work of composition and a long struggle to attribute responsibility to some actors.

34. C. Péguy (1914) is probably the most profound study on the articulations of the various historical and religious times. See also G. Deleuze (1968).

35. Apart from their respective know-how and professional loyalty, this is, the only distinction remaining between historians and sociologists of science: the former prefer starting from the temporal framework inside which the actors are situated, whereas the latter like to obtain the temporal framework as a consequence of the actor's movements. For the rest, both groups are doing the same job and are no longer separated by the absurd divide between empiricists, inter-

ested in details and narratives, and theorists, interested in structures and atemporal schemes.

36. See esp. M. Augé (1975), J. Favret-Saada (1977/1980). The process of accusation is an excellent model for the study of sciences as well as parasciences or witchcraft. By following *who* is preferably accused and *what* is preferably considered to be the cause of a misfortune, the ethnographer can easily reconstruct society's network of associations. Trailing the processes of accusation allows a direct entry into "sociologics." See B. Latour (1987, ch. 5).

37. This is why explaining Pasteur's success in terms of his ability to manipulate others, or in terms of his power over the hygienists, is so meaningless. If anything, Pasteur is the one who is manipulated from the start by hygienists in search of a solution to the conflict between health and wealth. But "manipulation" is a term like "power" or "strategy." All imply some degree of potency and are thus reductionist in essence (1.5.4.).

2. You Will Be *Pasteurs* of Microbes

1. Only if we distinguish between context and content does it appear contradictory to reduce the power attributed to a few great men and at the same time to highlight their personal contribution. The renewal by Tolstoy of the historical novel genre is a beautiful escape from this apparent contradiction: only after the crowds are put back into the picture can the novelist afford *each* individual his or her own flesh and color. Only when sociology has caught up with Tolstoy can we again be proud of our craft.

2. The word "strategy" is always used here in its *War and Peace* sense. That is, the strategist make plans that are constantly drifting away; he seizes upon opportunities in the midst of confusing circumstances; he fights hard to make others attribute responsibility for the whole movement to him in case of victory, while leaving it to someone else in case of defeat. This is no reason, however, for reducing action to microcontingencies and for appealing constantly to disorder, uncertainties, and idiosyncrasies. (K Knorr 1981, 1985). Each actor described by Tolstoy is *summing up* what the others do and is trying to make sense of chaos. Sometimes his interpretation is shared by others acting performatively on the setting, thus adding to the overall chaos. I call this performative summing up and negotiation of a global direction "strategy."

3. For Claude Bernard, see W. Coleman (1985). In spite of Coleman's renewed profession of faith in a bizarre dichotomy between "cognitive" factors and "social" ones, his article is, as usual, remarkable. Bernard makes a perfect contrast with Pasteur as far as the positioning of the laboratory is concerned. "Bernard's unswerving dedication to disciplinary limitation" (p.55) is precisely opposite to Pasteur's tactics of never discussing discipline boundaries and always crossing them. Moreover, Bernard places the laboratory in *juxtaposition* with hospital wards and physician's cabinets, expecting physiology, through a slow trickle-down effect, to influence practical medicine. For Bernard a laboratory is the "sanctuary of science"; for Pasteur it is a fulcrum and an obligatory passage point. Of course, they both consider an autonomous and well-funded science the fountainhead of everything else, but in my terms Bernard puts this autonomy at the level of the primary mechanism, whereas Pasteur puts it only at the level of the secondary mechanism. Coleman takes as real the distinction between cognitive

factors and social factors, which Bernard regards as one possible tactic for achieving autonomy. Had Coleman studied Pasteur, this clean distinction would have been developed in an entirely different way.

4. On the absurdity of such a link in the eyes of a late nineteenth-century physician, see J. Léonard (1977, 1981, 1986).

5. Once again, whenever I use the words "interest" and "interested," I am not referring to the "interest theory" expounded by what is now called the Edinburgh School. B. Barnes (1974); D. Bloor (1976). I am rather referring to the notion of translation. M. Callon, (1980). "Interest" means simply what is placed "in between" some actor and its achievements. I do not suppose that interests are stable or that groups can be endowed with explicit goals. On the contrary, I started from the notion that we do not know what social groups exist and that these groups do not know what they want. However, this ignorance does not mean that actors are not constantly defining boundaries, attributing interests, endowing others with goals, and defining what everyone should want. Any historical case study is thus an *in vivo* experiment in defining what the groups are, what they want, and how far we can negotiate with them. Interests cannot explain science and society; they are what will be explained once the experiment is over.

6. At this point it is crucial to treat nature and society symmetrically and to suspend our belief in a distinction between natural and social actors. Without this symmetry it is impossible to grasp that there is a *history* of nonhuman as well as human actors (see Part Two, sec. 3.0.0). The only way to understand this central part of the argument is to stick firmly to the semiotic definition of all actors, including the nonhuman ones. What is a microbe? An actor, that *does* this and that, in the narrative. Every time we modify one of the actions, we redefine the competence and the performance of the actor. This is how the *story* can show the *history* of the actors.

7. This reorganization of hygiene is misinterpreted even by an observer as meticulous as E. Ackerknecht (1967). Citing the same Bouchardat, he writes "The anticontagionism of our hygiene movement is probably one of the reasons why it has been so completely forgotten. After the sun of bacteriology had risen so high, the hygienists' anticontagionism looked a little embarrassing, and the whole movement receded into the shadows of insignificance . . . Belonging, like its clinical counterpart, to the prelaboratory era, the Paris hygiene movement of our period looked rather clumsy and stupid to the young enthusiast of the bacteriologist era" (p.160). The "rising sun" is one of those many metaphors historians like to use as a stopgap wherever the crucial question of the composition of time is at stake. Ackerknecht's interpretation is inaccurate. On the contrary, the notion of a "variation of virulence" allows hygienists to force enthusiastic bacteriologists to do *their* work ("their" being deliberately ambiguous). The fact that hygienists are ignored has nothing to do with success; it is a consequence of the secondary mechanism that the hygienists needed to employ in order to achieve their results faster.

8. In spite of Pasteur's importance, there are surprisingly few books on him. Apart from the hagiographic piece by R. Valléry-Radot (1911) and the moving book by Duclaux (1896/1920), there are only R. Dubos (1950) and an epistemological rendering by F.Dagognet (1967). For the Pasteurians, see Salomon-Bayet, ed. (1986). The only biography done by a professional historian is G. Geison (1974).

9. Here again the contrast with Claude Bernard's movement is striking. Pasteur is completely indifferent to disciplinary boundaries and to professional autonomy. See also F. Holmes (1974).

10. On Pasteur's passage from studies on molecular dissymmetry to "life sciences," see D. Kottler (1978).

11. This is the only instance in which the Tolstoyan meaning of strategy is replaced by the word's classic sense of an action successfully planned. The consequent steps that Pasteur is going to take are explicit in his correspondence and articles. There is no reason to abstain from recognizing that sometimes for a few moments there are indeed strategies. After all, even during the battle of Tarutino one or two columns arrived at the prescribed time and place (although not for the expected reasons).

12. Claude Bernard also recruits allies but in the opposite way. He insists on a precise order of command from science to practical applications *before* commencing the negotiation. See W. Coleman (1985).

13. As is well known, the French love revolutions. Time being seen as having no progressive and formative value, the only way to understand change is to imagine sudden breaks that transform one old regime into a new one. F. Furet (1978/1981) has shown the pregnancy of this myth for political revolutions. But it is much more powerful in the French history of science, which resounds with "epistemological ruptures" in Bachelard's, Althusser's, and Foucault's writings. A revolution, however is always the belated outcome of an attribution process and takes place only at the level of the secondary mechanism.

14. See R. Dubos (1950) and, for the French case, I . Grellet et C. Kruse (1983).

15. This is the main limitation of laboratory studies, including my own. K. Knorr (1981); B. Latour and S. Woolgar (1979/1986). They start out from a place without asking if this place has any relevance at all and without describing how it becomes relevant. In only a very few cases are laboratories the place to *start* with if we wish to see science in the making. Most of the time labs are dead ends, with everything interesting happening outside. For the dislocation of a laboratory, see M. Callon (1980). For the prehistory of another laboratory, see C. Salomon-Bayet (1978).

16. On this essential point a substantive body of literature has emerged since B. Latour and S. Woolgar (1979/1986). More and more scholars are becoming interested in inscription devices, instruments, visualization procedures, and other re-representation processes. See B. Latour and J. de Noblet, eds. (1985). On the medical aspect, see e.g. S. J. Reiser (1978).

17. G. Canguilhem (1977). If "the science of the laboratory was *of itself* directly at grips with the technical activity," the work of planning research and development would be an easy one (p.73). Epistemological definitions of the laboratory are no more relevant than sociological ones. It all depends on the earlier translations that render the "science" relevant to be the "technical activity."

18. This is why we do not have to choose between the two questions. "Has Pasteur discovered the microbes which were out there?" or "Has Pasteur socially constructed them?" The activity of discovering something is the same as that of commanding a network of equivalences. In this sense Pasteur has discovered his microbes just as Edison did his electricity. See Hughes (1979). That is, microbes and electricity were not much at first. It is only when they *added* as many attributes

as were necessary to interest everyone and to render their laboratories indispensable to the microbes and electricity, and only when they fought like devils to win attribution trials, that Pasteur and Edison ended up having discovered something.

19. Here again the definition of a new object is provided by semiotics. If we change the performances of any actor in the narrative, we modify its competence. In more ontological terms, since a shape is the front line of a trial of strength (1.1.6), if we modify one of these trials, we modify the shape. The name ("microbe," "bacillus") will correspond to a thing only when the front line has been stabilized. On this principle, see B. Latour (1987, ch. 2.)

20. A discovery is always retrospective and depends on the control of a translation network. Only if we pay this price do sentences like "what we thought until now to be anthrax is really caused by a bacillus" acquire some credence. If there were the smallest gap in the control of the translation, then Pasteur's "discovery" would simply be *added* to the complicated anthrax affair instead of *replacing* the old knowledge.

21. That there is a history of the "things in themselves" seems absurd only to those who want to fix us forever into the boring confrontation between a subject (or a society) and an object (or a nature). Meanwhile, innovators are constantly crossing the boundaries between nature and society and turning our careful distinction between what has been revealed, what has been discovered, what has been invented, what has been constructed, what has been made up, and what has been fabricated into a shambles.

22. As noted by M. Callon (1986), there should be a complete symmetry between the terms used to describe human and nonhuman actors. The first choice of term does not matter, but once we have chosen one for human actors, we shall stick to it when we address the nonhuman actors. If we "negotiate" with the microbes, then use the word for the hygienists or the ministry. If we "discover" bacilli, then "discover" the physicians or their colleagues. When this rule of method is applied, we soon realize that the distinction between science and society is an artifact caused by an assymmetrical treatment of human and nonhuman actors. The marvelous study of S. Shapin and S. Schaffer (1985) provides the genealogy of this distinction.

23. See B. Latour and J. de Noblet. (1985). See also F. Dagognet (1969, 1973, 1984).

24. For a "social construction" analysis of discovery, see A. Brannigan (1981). I am following here an "*associo*logical" analysis that relates the degree of "discovery" to the extension of a network. In this view Pasteur "discovers" microbes in the same way that electricity replaced gaslight. See T. Hughes (1983).

25. I see no reason to shun the term "genius." Only those who want to reduce the individual to the mass may object to this word. Such a reduction, however, would be unfaithful to Tolstoy's model. In his model no one is reduced to anyone else. Those able to sum up, locally and for a time, what the others do should be admired without reservation. This is what Tolstoy does with Kutuzov and what I do here with Pasteur's primary mechanism.

26. According to D. Watkins (1984), there is a difference between French and English professionalization strategy in this respect. The possible short cut between basic science and medical practice is much more pronounced in France than in

England, where a new profession arises, preventive medicine. See also W. Frazer (1970).

27. According to Nicol (1974), among the precautions to be taken were the shaking of the flasks of vaccine and the injection of one control and one vaccinated sheep from the *same syringe* so that Pasteur could not be accused of cheating by injecting virulent forms to the "nonvaccinated" and attenuated forms to the "vaccinated" (p.377). The negotiation was serious because Hippolyte Rossignol, who organized the challenge, explicitly set it up to disprove Pasteur's claims and to show him "that the Tarpeian rock is close to the Capitol" (p.368). Founder of the Société de Médecine Vétérinaire Pratique, Rossignol organized the public experiment in part as a publicity stunt for his journal, *La Presse Vétérinaire*.

28. But the anthrax vaccine crosses the Hungarian border like a bullet. "The Hungarians," writes Thuilier to Pasteur on October 1, 1881, "are even greater admirers of your discovery than I had thought at first. They are firmly convinced of its truth. The demonstration experiments that I am performing are actually of only moderate interest to them—they are so convinced in advance of success. What interests them more is to know (1) how to prepare pure cultures, (2) how to make the vaccine." Pasteur and Thuilier (1980), p.91. Good network builders, these Hungarians. They even try to corrupt the young Thuilier so that he reproduces in front of them all the gestures necessary to turn the vaccine into a reality.

29. Like the notion of discovery, that of an application of science "outside" is an artifact obtained once the activity of network building is over. Instead of limiting ourselves to social construction and denying that microbes are out there and have been discovered, we simply have to give qualified answers to these questions; the qualification consists merely of *adding* the activity of network building. The distribution of the microbes "throughout the world" is exactly similar to that of gas and electricity.

30. I am limiting myself to the article, but a full account of the episode is found in H. Mollaret and J.Brossolet (1984). They make much of the priority dispute with Kitasato but also show clearly the contrast between the Pasteurian strategy and that of the English, the Chinese, or the Japanese physicians and biologists.

31. On the French debates around water and sewages, see Goubert J.-P. (1985), G. Dupuy and G. Knaebel (1979); A. Guillerme (1982).

32. H. Mollaret and J. Brossolet (1984): "Whereas, schematically speaking, Koch and his school tried to identify the agents responsible for human and animal infections, Pasteur and his disciples tried to attenuate the pathogenic power of these same agents to turn them into vaccines." (p.150).

33. As shown by N. Jewson (1976), this renewal had been taking place for more than a century. Before the advent of hospital and laboratory medicine, Jewson argues, "It was the sick person who decided upon the efficacy of his cure and the suitability of the practitioner. Hence practitioners, and thus medical investigators, formulated their definition of illness so as to accord with the expectations of their clients" (p.232). The history of medicine, then, is the history of the reversal of this dependence upon the client and the sick person. In this sense Pasteurian definition adds still more distance to the estrangement from the sick.

34. Pasteur during this period has discovered not "the microbe" but the mi-

crobe-that-can-be-attenuated, and this actor existed from the early 1880s to the end of the century. That is why the notion of discovery is so useless. It can work only in the temporary period of calm on the front line. As soon as the struggle starts up again, the objects have new properties. See L. Fleck (1935/1979).

35. For the United States, see R. Kohler (1982).

3. Medicine at Last

1. See J.-F de Raymond (1982) on the first vaccination. The story has many aspects similar to Pasteur's. It is tied to state intervention and statistics. Jenner slightly transforms an earlier practice (innoculation). Even the "*associo*logy" works along similar lines. "Immunology allows one to dispense with an ethnic or a social segregation" (p.111). On French first vaccinations, see P. Darmon (1982).

2. For this sort of reason we cannot even assume that 1892 is before 1893. It could as well be "after," or "at the same time." It all depends on what actors do to place these years in relation to one another (1.2.5)

3. See A.-L. Shapiro (1980). The more the law was discussed, the more it was "emasculated" from the hygienists' point of view and the more it maintained the traditional position of physicians. On the medicolegal history of this period, see R. Carvais (1986).

4. This situation is not limited to France. For the United States, J. Duffy (1979) writes about the declaration of tuberculosis: "The intimate relationship between the physicians and the patient's family in the upper class and the danger of losing his fee among the lower economic groups tended to discourage reporting disease which might have serious economic consequences to the family" (p.10).

5. This specific kind of health officials, called "officiers de santé," were doctors without the national academic diplomas but with some kind of legal protection as a consequence of the French Revolution's movement to dissolve entirely the "privileges" of the medical profession. For a century each issue of each medical journal attacked the existence of these inferior "officiers de santé" who took the bread out of the real doctors' mouths. On the complicated French legal scene, see M. Ramsey (1984); R. Carvais (1986). On the problem of professionalization in the medical profession, see E. Freidson (1970).

6. "With cries of approval from the right, M. Volland prophesied in the Senate that: 'By the law of hygiene that you consider today, you will have armed the representatives of the central power with the right to penetrate when they wish, on an order from Paris, day or night, inside our homes; to bring, in defiance of all the guarantees laid down by the criminal code, into our homes their war on microbes, and under the pretext of the search for a germ or the execution of a disinfection, to open our most intimate possessions and our most secret drawers.' " Cited in A.-L. Shapiro (1980), p.17.

7. D. Watkins (1984) reports of the professionalization in London, which, if it can be extended to all of England, makes a striking contrast with the French case. "Poor law medical officers, though employed in public service, continued to practice curative, clinical medicine, in the same way as their private practitioner colleagues. Medical officers of health however were practicing a different type of medicine altogether. The function of their office required specific training in a specialized area of knowledge. This specialized practice begat its own aims, goals, and objectives. Consolidation of these through the professionalization of

preventive medicine resulted in a sub-division of this occupation from the medical profession as a whole" (p.16). See also W. Frazer (1950).

8. For this notion of a deal or a contract in the French medical profession, see e.g. C. Herzlich (1982): "Simultaneously, in what can be called an 'exchange-process,' physicians let it be understood that they would co-operate with the social laws and 'enter into the social game' of collective relations, but only under certain conditions which they were able to impose in exchange for their co-operation, and which shaped medical practice" (p.245).

9. The situation is the same for doctors as for the hygienists a generation earlier. We need a "translation platform," so to speak, that is ambiguous enough to aggregate interests. Contagion does not interest hygienists; variation of virulence does interest them, because it resembles what they were already doing and allows them to fuse the macrocosm—the city—and the microcosm—the bacilli cultures. Vaccines do not interest physicians very much; sera interest them a lot, because they can go on doing their usual work. In both cases the price to pay is the same: laboratory equipment. In both cases the Pasteurians are the ones who modify their angle of attack and their research program. The variation of virulence was not comprised at first in the earlier definition of the microbe. As for the serum, it was not part of the research program before the constitution of immunology. Vaccines and sera are thus a *coproduction* of the Pasteurians, their human allies, and their nonhuman captive allies

10. On this point, G. Weisz (1980) shows that Pasteurism does not play a very important role in the transformation of French medical education. More important is the contract made between physicians and the state: "eliminate our competitors and we will become more knowledgeable." The influence of Germany and its creation of a full-time teacher-researcher also play an important part (p.64). Among the chairs created between 1870 and 1919, very few are in the "pasteurized" domains. In general, the whole linkage between science and medicine is considered uncertain and often unnecessary by students and general practitioners. See Salomon-Bayet (1986).

11. There are times when sociological notions, such as that of "prestige," can be used. Such is the case in this chapter on physicians, because we are now much further from the technical content of bacteriology and are talking about a group that has become the epitome of a profession. See Friedson (1970); Starr (1982). The further we are from content, the better traditional sociology is.

12. What happened to Villermé and the hygienists happens now to the pasteurized public health. They both start as a new science in search of allies; they both end as a reified social movement that has aggregated so many people along so many networks that notions of power appear applicable.

13. L. Murard and P. Zylberman (1984) criticize this point, rightly so from their point of view. It is true that later in the century hygiene is metamorphosed many more times, especially because in the long run the alliance between hygienists and politicians does not work very well. The notion of "sanitary police" becomes embarrassing. Still, in contrast to their importance in the earlier period, the themes of hygiene disappear and become routinized.

14. McNeill (1976); F. Cartwright (1972); M. Worboys (1976).

15. The extension of micro- and macro-parasites is especially striking because, as J.-P. Dozon (1985) argues, many of the diseases were new ones imported by the very columns in charge of eradicating disease.

16. In effect the Pasteurians resolved the conflict between Manson's and Ross's

approaches that are illustrated by M. Worboys (1976): "The difference [between the two scientists] came over whether it was to be 'scientific research' for development, or 'public health' for development" (p.91).

4. Transition

1. Those who accuse relativists of being self-contradictory (Isambert, 1985) can save their breath for better occasions. I explicitly put my own account in the same category as those accounts I have studied without asking for any privilege. This approach seems self-defeating only to those who believe that the fate of an intepretation is tied to the existence of a safe metalinguistic level. Since this belief is precisely what I deny, the reception of my own argument exemplifies my point: no metalinguistic level is required to analyze, argue, explain, decide, or tell stories. Everything depends on what sort of actions I take to convince others. This reflexive position is the only one that is not self-contradictory (Latour: 1988).

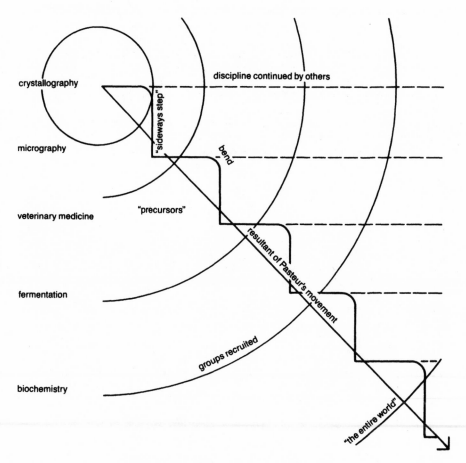

Figure 1. Pasteur's trajectory (see p. 69)

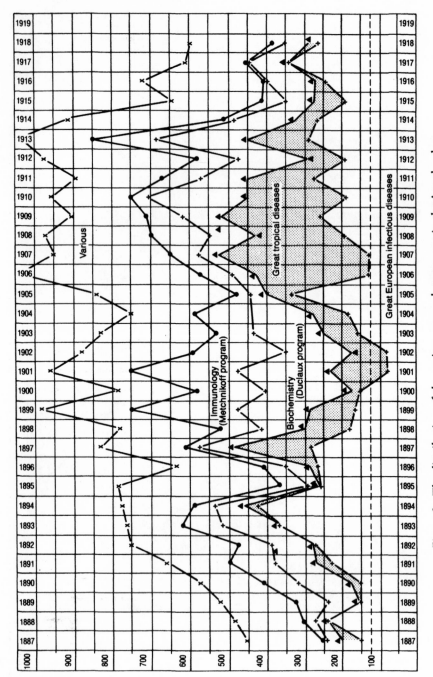

Figure 2. *The distribution of the various research programs in the* Annales de l'Institut Pasteur *(by number of pages), 1887–1919 (see p. 101)*

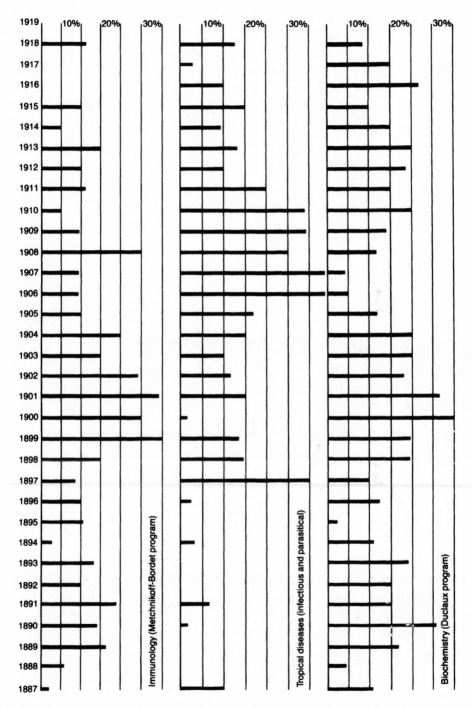

Figure 3. The variation of the three most important research programs in the Annales de l'Institut Pasteur *(by year and percentage), 1887–1919 (see p. 106)*

p13 Do we need to know of Henrietta

p15 There must be more than one actor

Henrietta Honored —
why not George Gey?

p16 An idea never moves of its own accord

p18 Conflict between health + wealth

p19 Social Context of Science

p21 To understand everything is
to understand nothing

p22 If we say man has moved a mtn

p32 Must take everything into
account @ mce

p36 everyone linked due to fear

p37 Note #25
Interdependence

p40 Science of Associations

p41 Health + Wealth

p117 Never foreign patent

p122 No more diseases

p131 Evolution/Revolution
Precious victories available to all

Index

CDC
visit
May 13, 2013

Computer room | Phone line to Teacher on the front line | |

Emergency Operations Center
for curing Education

Poet
Whitman

Explorations
Expert
Humboldt

Role of Each Scientists

Evolution   Taxonomist
Microbiologist
Cell Biologist

26
27

91-92
92-93
93-94